THE QUANTUM DIVIDE

T0202399

THE QUANTUM DIVIDE

DIVIDE

Why Schrödinger's Cat is Either
Dead or Alive

*Christopher C. Gerry
and Kimberley M. Bruno*

OXFORD
UNIVERSITY PRESS

OXFORD
UNIVERSITY PRESS

Great Clarendon Street, Oxford, OX2 6DP,
United Kingdom

Oxford University Press is a department of the University of Oxford.
It furthers the University's objective of excellence in research, scholarship,
and education by publishing worldwide. Oxford is a registered trade mark of
Oxford University Press in the UK and in certain other countries

© Christopher C. Gerry and Kimberley M. Bruno 2013

The moral rights of the authors have been asserted

First published 2013
Reprinted with corrections 2013
First published in paperback 2016

Published in the United States of America by Oxford University Press
198 Madison Avenue, New York, NY 10016, United States of America

British Library Cataloguing in Publication Data
Data available

Library of Congress Cataloging in Publication Data
Data available

ISBN 978–0–19–966656–0 (Hbk.)
ISBN 978–0–19–875407–7 (Pbk.)

Cover image: © shutterstock.com

Dedication

CCG: To the memory of my parents, Clayton and Phyllis Gerry

KMB: To my parents, Paul and Mary Ann Bruno, for your unending love and support

Preface

This book is about the essential ideas of quantum physics as elucidated by a selection of key experiments, mostly, but not entirely, taken from the field of quantum optics, the study of the interaction of light and matter. The intended audience for this book is anyone with a keen interest in learning about nature of the quantum world as revealed by intriguing experiments performed over the past few years. This includes layman as well as students of physics.

In this book, we discuss a number of experiments chosen to illustrate the sharp discontinuity in the way one thinks about physical phenomena in the everyday world of the large scale and the way nature forces us to think about phenomena that occur on the scale of the very small, the scale of atoms. It isn't just that atomic-scale phenomena are very different than large scale phenomena, but that the former seem to not conform to the logic of the large-scale phenomena of everyday experience. The planet Mars is right now in a particular location on its orbit around the sun. We don't know what it is at this moment of writing, but we know we can find out easily enough. Even if we don't know the position of Mars, we can nevertheless be assured that it does have a definite location in space at any given moment. On the other hand, think of the simplest of atoms: the hydrogen atom which consists of a single proton and a single electron held together by the electrical force of attraction between them. In the simplest quantum-like model of the hydrogen atom, the so-called Bohr model of 1913 that you probably encountered in your high school chemistry course, the electron orbits the more massive proton very much like the planets orbit about the sun. However, in the modern version of quantum mechanics, developed in 1925-26, an electron doesn't have such easily visualized electron orbits. In fact, it doesn't have *any* orbits at all in the ordinary sense of that word: it has only a probability distribution in the space around the proton. Furthermore, the quantum theory allows for situations where the electron could be in a special kind of state where it *superficially* seems to be on *both* sides of the atom at the same time. We hasten to emphasize that quantum mechanics *does not* actually say that an electron can be two places at once, hence the use of the proviso that quantum mechanics only superficially appears to allow the electron to be in two places at once. Such a question would never even occur in the connection with the motions of large scale objects be they planets, baseballs, or grains of pollen. Of course, we have no direct experience with the atomic world. But strange states of matter and of light can and are produced routinely in laboratories around the world. As we indicated above, it is not quite accurate to say that even a quantum particle can be in two places at once. Things are much more subtle than that. We shall also ponder the prospect that some of these very weird atomic scale quantum phenomena can actually make an appearance in the everyday world. In fact, the main title of this book, *The Quantum Divide*, references that exact problem: *where one can draw the line between the classical and quantum worlds?* One possibility is that no such divide may actually exist.

We do *not* present quantum mechanics through a historical account of the development of the subject as there are numerous books already available for that purpose. However, certain historical references are unavoidable. As an aid to the reader, we provide, as an appendix, an historical outline (timeline) that highlights the primary developments of the subject, including relevant experiments, and several books that follow the historical development can be found in the bibliography. For the most part, we do not deal with the personalities of those involved with that development and interpretation of quantum mechanics, or those who continue to elucidate the strange nature of the quantum world in the laboratory and in theoretical studies. Again, numerous books have already appeared where history and personalities have been discussed to some degree or another, though sometimes at a superficial level. Indeed, sometimes quantum physics itself is described rather superficially in these books. Our intention is to stick to the physics of the quantum world, with the expectation that the world at that level has more than enough quirky and counter-intuitive phenomena to keep the reader intellectually challenged and at the same time even entertained.

In our presentation, we do not shy away from using some aspects of the mathematical formalism of quantum mechanics, particularly for the representation of quantum states and their superpositions, and for entangled quantum states. This is done to help the reader better understand (we hope!) what quantum theory is trying to tell us about the world. No actual calculations are performed in any of our discussions.

Acknowledgments

We first and foremost thank profusely Dr Jaroslav Albert for preparing all the figure drawings for the book using CorelDraw, and for doing one drawing by hand (Born's machine gun). CCG wishes to thank Rainer Grobe and Mark Hillery for many informative quantum conversations over the years. We thank all who have critically read various versions of the manuscript and have made valuable suggestions, almost all of which we have adopted. Any errors that remain are entirely our responsibility.

Contents

Physics Fundamentalism

A man said to the universe:
Sir, I exist!
However, replied the universe,
The fact does not create within me
a sense of obligation.
STEPHEN CRANE

1.1 Dividing Up the World

It has famously been said that there are two kinds of people in the world: those who divide the world into two kinds of people, and those who do not. But human beings have always had a preoccupation with classification. Despite our valiant efforts, divisions that we propose and categories that we form are often fluid. In science, we differentiate between the three basic natural sciences—physics, chemistry, and biology—in the sense that each must be studied in its own terms. However, we also fully realize that there is much overlap among them. Within physics, the proper category for this book, we find another kind of division, a divide which seems rather natural. First, there is the physics of the *macroscopic* world, that is, the world of everyday phenomena and of the universe on a large scale. This includes the motion of golf balls, planets, galaxies, clusters of galaxies, and so on. In this world, the laws of classical physics—the laws of Newton, Faraday, Maxwell, Einstein (his gravitational theory, we mean here), and so on—hold true. But then there is the physics of the *microscopic* world—the world of atoms, molecules, photons, quarks, and so on—which appears to be operating by a rather different set of laws. Although there are vestiges of the classical laws in quantum physics, the way we must think about the quantum world is very different than how we think about and describe mathematically the classical world.

Quantum mechanics is supposed to be a more fundamental theory than classical mechanics. Moreover, classical physics presumably emerges from quantum physics as a limiting case when certain parameters, (energy, momentum, and so on) become large. Nevertheless, there is a divide between the way the world works on the microscopic and macroscopic scales. An important question concerns the "location" of the quantum/classical divide: At what scale does quantum mechanics go over to classical mechanics in the sense that one can dispense with the strange outlook provided by the quantum theory and go on using the familiar outlook of everyday life? In addressing this question we must examine the *mesoscopic* world—the world between the two extremes where somehow the laws of both regimes should merge. At the same time, we must be sufficiently open-minded to consider the possibility that in a large context there really is no quantum/classical divide at all, that the perceived divide is an illusion. After all, it is quantum mechanics, not classical mechanics, that is generally thought to be the most fundamental theory. So perhaps the real question is not so much about the location of the quantum/classical divide rather than about whether there even is a divide.

As the reader will learn, the major difference between the classical and quantum domains is not just in the mathematical forms of the laws. In fact, we will not even explicitly write down these laws. Rather, it is the way we are forced to think about the quantum world and of what can objectively be known about it. To highlight the differences in the ways we need to think about the quantum world versus the classical world, let us consider the following example: Suppose you have a cat and you know with complete certainty that it is inside your house, but that you are outdoors so that you do not know *exactly* which room it is in. Nevertheless, being a rational person you would probably conclude that the cat is definitely *in one of* the rooms. We could call this *objective ignorance*: The cat is definitely somewhere, but we do not know where. This makes sense in the everyday, classical, world. The cat's location is *objectively definite* even if unknown. In very sharp contrast, if the cat were an object that obeyed the laws of quantum physics (which it does not, because it is macroscopic in scale) it could be the case that the cat is in the house but not in a definite location within the house. It would not be just a matter of your not knowing in which room the cat is located; rather, its location would be *objectively indefinite*. That is, it has no definite location at all. Such a shocking notion is never encountered in ordinary life. However, in the quantum world, electrons, photons, and other "microscopic" particles can, under certain conditions, have attributes that are objectively indefinite. Certainly, the idea that *anything*, cat or electron, may not have well-defined attributes is shocking when one first learns about it, and, we must admit, remains shocking even after many years of thinking about it. It is just "common sense" to posit that a cat, or any object, will be in either one place or another, and a rational person would not even think twice about it. But in the atomic world it is commonplace for objects to have objectively indefinite attributes. In such cases there is a kind of "smearing-out" of the attribute in question. For our example, the *location* of the cat could be smeared-out over the house, though we do not hesitate to say that the cat itself is *not* smeared over the house. Quantum theory does *not* predict that an object can be in two or more places at once. The false notion to the contrary often appears in the popular press,* but is due to a naïve interpretation of quantum mechanics. Nevertheless, the idea that objects can have attributes that are objectively indefinite certainly clashes with our everyday "common-sense" view of the world. And objective indefiniteness, however, is perhaps the least shocking thing about the quantum, as the reader will soon discover, but much of the rest of quantum weirdness follows from it. One thing that is worth keeping in mind as you read the following pages is that while quantum *theory* is weird, its weirdness is no more than a reflection of the weirdness of nature *itself* on the level of the atomic world. That is, experiments reveal counter-intuitive phenomena in nature that give rise to the modern quantum theory, which provides consistent explanations in the form of mathematical expressions of the laws of nature in the atomic world.

Before we enter into our subject proper, it would be useful to better understand the relationship of physics to the fundamental sciences.

---

* For example, the cover of the June 2005 issue of *Discover* magazine asks: "If an electron can be in two places at once, why can't you?" Well, quantum theory does not say that even an electron can be in two places at once.

1.2 Physics as Fundamental

We begin with the following pronouncement: Physics is the most fundamental of all the natural sciences. To anyone not a physicist (which means approximately everybody) such a proclamation needs justification.

According to the dictionary, physics is the science that deals with matter and energy and their interactions in the fields of mechanics, acoustics, optics, heat, electricity, magnetism, radiation, atomic structure, and nuclear and elementary particle phenomena. So, why do we think of physics as being somehow more fundamental than any of the other sciences?

Consider biology, the science of life. Fundamental to biology is deoxyribonucleic acid, commonly known as DNA, the molecule of life. DNA stores and transmits genetic information from one generation to the next. In other words, it provides all instructions for the creation and maintenance of life. Each trait of an organism is encoded in a segment of DNA called a *gene*. Each gene is constructed out of groups of other kinds of molecules called nucleotides, and nucleotides are constructed out of atoms of the elements carbon, hydrogen, oxygen, nitrogen, and phosphorous. The atoms of the various elements are more elementary than molecules. Biology may be ultimately *reduced* to chemistry—complicated chemistry, but chemistry nevertheless.

Chemistry, as we all know, is the study of how different chemical substances can combine to form new chemicals. Chemists spend a good deal of their professional lives thinking about atoms and how atoms bond together to build molecules. Therefore, underlying chemistry must be the science that describes the atoms themselves, and, as may be deduced from the dictionary definition above, this science to which we are referring is *physics*.

Although the atomic hypothesis had been proposed by John Dalton in the early 1800s, the rules governing the structure of atoms and the mechanisms of how they are able to bond together to form molecules were not worked out until the early part of the twentieth century. These rules were devised mainly by physicists applying the newly developed theory known as *quantum mechanics*. Not until there existed an understanding of the chemical bond could it be said that chemistry had a firm theoretical basis. Chemistry, in this sense, has been *reduced* to quantum physics. Of course, chemistry must be studied in its own terms, not merely as a branch of physics. Nevertheless, it must always be kept in mind that quantum physics underlies the entire discipline.

Physics, however, is not just about atoms, which are a relatively recent scientific discovery. It is fair to say that throughout most of the history of physics, atoms were hardly the center of attention. Once called natural philosophy, physics originated in the study of the motion of inanimate objects. From those studies emerged a body of knowledge, including various natural laws, that today we call "classical" physics. Newton's laws of motion, the basis of the science of mechanics, come to mind, as does his law of universal gravitation. These laws are obeyed by objects that in some sense are "large" and, in fact, the laws were discovered after careful observation of the behavior of such objects. But what do we mean by *large*? It is difficult to be precise. Perhaps the best way to delineate *large* objects from *small* objects is to operationally posit that objects obeying Newton's laws are large-scale objects. Since this seems like circular reasoning, let us explain further. Planets, obeying Newton's laws of motion and his law of universal gravitation, that are orbiting about the Sun, are clearly "large" in this sense. In fact, on a human scale they are quite large. However, a golf ball also obeys Newton's laws, as do much smaller particles such as specks of dust and grains of pollen. These objects are also thought of as "large" in this sense. Therefore, Newton's laws have a vast range of validity.

After the time of Newton, electricity and magnetism (*electromagnetism*) due to the work of Franklin, Faraday, Maxwell, and others, and *thermodynamics*, the study of heat and its transformation, which was stimulated by the development of steam engines during the industrial revolution, was systematically developed by Joule, Carnot, Clausius, and many others, and incorporated into what we now call *classical physics*. Again, this was done on the basis of observations of phenomena on a fairly large scale. For example, as demonstrated by Ben Franklin, lightning is a large-scale electrical phenomenon. So is the magnetic field that affects the needle of a compass. Thermodynamics is a theory of heat and its transfer that considers matter only in bulk and completely disregards underlying structure. It is therefore a *phenomenological* theory. A phenomenological theory is one that mathematically models phenomena without constructing a detailed microscopic picture underlying the phenomena in question. Thermodynamics does not concern itself with the details of the bulk systems that it describes. It is a powerful theory that allows us to construct useful devices such as refrigerators and internal combustion engines. The laws of electromagnetism and of thermodynamics were established during the eighteenth and nineteenth centuries.

In the late nineteenth century and into the early twentieth century, it was recognized that something working on a much smaller scale seemed to underlie electromagnetism and thermodynamics. As we now understand, electricity and magnetism depend on the fact that there exist tiny particles that carry electric charge and act as if they also carry tiny circular electric currents that generate tiny magnetic fields. Most everyday electromagnetic phenomena can be explained in terms of the motion of the charged particles that we call *electrons*. An electric current is nothing but the flow of charged particles—usually electrons—and electrical currents generate magnetic fields that surround the currents. If a conducting wire is placed in a *changing* magnetic field, or moved within a magnetic field, a current will flow within the wire. This is the basis of modern large-scale generation of electricity. But think now of a so-called "permanent magnet". There are no large-scale electrical currents flowing in the slab of iron or nickel that constitutes a permanent magnet. The origin of this magnetism is *atomic*. Specifically, electrons act as though they are spinning and carrying tiny electric currents inside them—quite a feat for particles that have no physical size whatsoever, as far as any experiment has been able to show. Nevertheless, it is ultimately the *combined* effects of many such atoms that are responsible, when they are all properly aligned, for the magnetic field of the entire permanent magnet.

Thermodynamics, it turns out, can be reduced to *statistical mechanics*—a theory developed by Maxwell, Boltzmann, Gibbs, and many others—which relates averages taken over the motions of small particles to bulk properties such as pressure, heat capacity, and so on. Heat is now fundamentally understood in terms of the average energy of motion of these small particles. The higher the temperature, the more energy the particles have.

Atoms have been found to be made up of even smaller particles—namely electrons (mentioned already), protons, and neutrons. Electrons carry the negative charge and protons the positive, while, as the name indicates, the neutron is electrically neutral. But the trend of finding structure on a smaller scale has continued. It now seems that protons and neutrons are made up of other particles—the quarks and the "gluons"—that mediate the strong nuclear force binding the quarks together. (Quarks have never been seen as "free" particles; they are thought to be permanently trapped inside protons and neutrons and other particles that interact through the strong nuclear force, although this has yet to be proved from the theory of the strong nuclear force.) At this point we have reached the level of our present understanding of the structure of matter. There has been speculation that quarks might be composed of something even more elementary, but as yet there is no compelling experimental evidence that this is so. Note that reductionism takes place at every stage: the properties of matter on one scale depend on the properties of matter on a smaller scale (and on fields through which the particles of matter interact), and so on, to the next smallest scale. It is this modern, and secular, version of the "Great Chain of Being" that informs us of the functional relationships between the small world of atoms and of the macroscopic world of everyday life.

This book mostly is about the "small" world of the atom, by which we mean not only atoms, but also electrons, photons, atomic nuclei, molecules, and even some aspects of solids. In that world, as we have said, the laws of physics turn out to be different from the laws of classical physics that we use to explain phenomena in the macroscopic world of everyday life. Although vestiges of classical physics persist, those laws just do not work in the atomic domain. Perhaps it is more accurate to say, given that quantum mechanics is the more fundamental theory, vestiges of the *quantum* disappear as we go into the classical world. Early attempts at providing a quantum theory, from 1900 to 1925, did, in fact, attempt to modify classical laws with various *ad hoc* rules. But these rules, already *ad hoc*, seemed to need *ad hoc* modification for almost every new application, and in some cases, no rules whatsoever could be found to explain observations. The quantum theory of this period, now known as the *old quantum theory*, pretty much ran out of steam by the early 1920s, and was in any case unsatisfactory due to the fact that the rules for quantization could not be determined systematically from any general principle. In 1925, new laws of the physics of the atomic world were discovered independently by Werner Heisenberg and Erwin Schrödinger. This new set of laws is collectively known as *quantum mechanics*. So, we now have two sets of laws: one set operating in the world of the *large*, the world of *classical physics*, and another operating in the world of the *small*, the world of *quantum physics*. In fact, there is a great divide between these two worlds, not only with respect to phenomena and the laws operating on the two levels, but also with respect to the way one must think about them. Classical intuition and the "common sense" of everyday life do not apply in the quantum world. Perhaps the most startling difference between these two worlds is with respect to the issue of causality—the principle that events are always preceded by their causes. In the classical world, when something happens it happens for a reason. But in the quantum world events can occur without any reason. The theory gives us only statistical predictions, i.e. the probabilities for events to occur, but gives us no deeper picture with regard to causes. The statistical predictions of quantum mechanics are inherent. There is nothing quite like this in classical physics.

Our goal in this book is to demonstrate and elaborate upon the failure of classical intuition through a discussion of a number of mostly recent experiments on quantum mechanical foundations, most of them involving light at the level of a few photons—sometimes only one. The relevant area of research is known as *quantum optics*—the study of the nature of light and its interaction with matter. This is an important area of contemporary physics research, not only for its intrinsic interest in elucidating the nature of light and its interactions with matter, but because of its potential applications to the emerging field known as *quantum information processing*. This field includes quantum computing and quantum cryptography (otherwise known as quantum key distribution)—issues that we touch upon in the later chapters. Among many things, we shall address the question as to whether or not there is any possibility that quantum phenomena can, on occasion, cross over into the world of the large, even if only briefly.

The quantum world differs in several ways from the picture we have of the classical world. One significant difference has to do with the role of measurement. In the latter case, if we measure, say, the position of a planet or of a baseball, the measurement itself is not suspected of having any effect on the future motions of such objects. We assume further that when we measure the position of a baseball, one that is just sitting on the field, to make it simple, that the measurement merely reveals the position it had just before the measurement. However, as we shall see, in the quantum world measurements generally do not reveal pre-existing information about quantum systems, and because of this they can be used to steer quantum systems into useful states for practical applications.

We cannot experience the quantum world directly through the senses. However, carefully controlled experiments have shown that the quantum world holds more surprises than the famous rabbit-hole of Alice in Wonderland. It turns out that in the quantum domain, the experimenter *can*, in very subtle and very limited ways, influence the *kinds* of results that can be obtained from a complete experiment. By making certain kinds of choices in experimental design, nature can be forced to behave in certain mutually exclusive ways. Other choices force a different kind of behavior, a complementary behavior, with qualitatively different *kinds* of results. In general, the results of any given run of an experiment cannot be predicted, as the quantum world is not deterministic and many of the predictions of the theory are statistical, as we shall explain in the pages that follow. Unfortunately, this aspect of quantum mechanics has led to all kinds of distortion and hyperbole in the popular press, some of which we address in the final chapter. In the balance of the book we intend our presentation to be as sober as possible and to let the facts stand for themselves. Quantum phenomena are strange enough on account of their contradictions with common sense without any need for hyperbole. You, the reader, will find in the course of reading this book that "common sense" (highly overrated in science anyway) may be almost entirely (but not totally!) tossed to the wind when it comes to quantum phenomena. That is what makes the quantum world so fascinating.

Bibliography

A few books on the history of classical physics.

Bernal J. D., *A History of Classical Physics: From Antiquity to the Quantum*, Barnes and Noble, 1997.

Newton R. G., *From Clockwork to Crapshoot: A History of Physics*, Belknap Press, 2007.

Segrè E., *From Falling Bodies to Radio Waves*, Dover, 2007.

The Duality of Particles and Waves: The Split Personality of Electrons

2.1 The Macroworld versus the Microworld

> *How often have I said to you that when you have eliminated the impossible, whatever remains, however improbable, must be the truth?*
>
> ARTHUR CONAN DOYLE, *Sherlock Holmes: The Sign of Four*

We hope you have been convinced by the discussion of the previous chapter that physics is *the* fundamental science. If so, you might expect that the foundations of physics are rock solid. Well, they are and they are not. Large-scale phenomena are well described by the classical laws of physics, mostly established by the beginning of the twentieth century, as we discussed in Chapter 1. But by the early 1900s it was realized that these laws *do not* seem to work very well when applied on the atomic scale. In 1911 Ernest Rutherford discovered the atomic nucleus, wherein resides most of the mass of an atom and all of its positive charge. He found that the nucleus contains all of the protons (the positive charge in an atom) and the neutrons (chargeless particles), but takes up only a very small volume of the atom. He did this by bombarding a gold foil with alpha particles (now known to be the helium nuclei—two protons and two neutrons), and noticed that occasionally the alpha particles scattered straight backwards—a feat not possible if the positive and negative charge of the atom were distributed uniformly throughout, as was thought to be the case at the time. But if the positive charge of the atom were concentrated in a massive nucleus, the occasional backscattering of the alpha particles could be explained as being the result of near head-on collisions of alpha particles with a nucleus. Furthermore, knowledge of the energies of the alpha particles allowed for the determination of the closest approach of the particles to the nucleus, which in turn gave an estimate of the size of the nucleus. So, Rutherford discovered not only the existence of the atomic nucleus, but also its size.

A sense of the size of atomic particles was available long before the work of Rutherford. In an attempt to "still the waves", in 1771 Benjamin Franklin poured a teaspoonful of oil on a pond in Clapham, England, and noticed that the oil spread over a surface area of about half an acre (2,000 square meters = 2×10^3 m^2). For a *volume* of oil of 2 cubic centimeters (or 2×10^{-6} m^3), and with the assumption that the layer of oil on the pond is about the thickness of one molecule of oil, one can divide the volume of the oil by the area of the spread and obtain an order of magnitude* estimate of the thickness d of an oil molecule: $d = (2 \times 10^{-6}$ m$^3)/(2 \times 10^3$ m$^3) = 10^{-9}$ m. (We have used scientific notation, where $100 = 10^2$, $1000 = 10^3$, etc.; $0.1 = 1/10 = 10^{-1}$, $0.01 = 1/10^2 = 10^{-2}$, etc.).

Now 10^{-9} m $= 0.000\ 000\ 001$ m is a unit of length denoted as one *nanometer* (nm), which is of the correct order of magnitude of the dimensions of a molecule. Alas, Franklin did not recognize the significance of his discovery; it appears he was thinking only of the thickness of the oil-slick and not of molecules. He noticed that the oil had the effect of reducing the wave action of the water, and considered the application of oil-on-water to reduce wave action on ships at sea. Unfortunately, only the small capillary waves are reduced by the oil, not the large destructive waves.

Now an atom is about one order of magnitude smaller than the smallest molecules, and thus has a diameter on the order of 10^{-10} m ($0.000\ 000\ 000\ 1$ of a meter). But Rutherford's experiments placed an upper limit of about 10^{-15} m —a unit known as a *femtometer*—for the size of an atomic nucleus. One consequence of this discovery of Rutherford's is the astounding realization that ordinary matter composed of atoms is mostly empty space. How so? Consider the following. The atomic nucleus is about five orders of magnitude smaller than the size of the atom itself. The electrons are apparently

---

* The phrase "order of magnitude" refers to powers of 10, but in an approximate sense. A factor 10 is one order of magnitude, a factor 10^2 is two orders of magnitude, and so on. The number 11 is approximately 10, so we say that is of the order magnitude 10, whereas the number 98 is approximately 10^2, so we say that the number 98 is an order of magnitude greater than 11. The actual numerical values of the numbers are not important; only their approximate relative sizes given as powers of 10 are of interest.

point-like particles and take up no volume at all. To obtain a sense of the relative sizes of an atom and its nucleus, suppose we scale up everything by a factor of 10^{13}. This scaling will render the nucleus a diameter of 10^{-2} m $= 1$ centimeter—about the size of a marble—whereas the atomic diameter scales to 10^{3} m $= 1$ kilometer. The electrons themselves do not scale up in size at all; they remain point particles. So if we imagine an atom to have a diameter of 1 km, and that the nucleus is about the size of a marble, we obtain a sense of how much of an atom is really empty space. Thus, the apparent solidity of, say, a baseball and a bat, must be just an illusion. The fact that the bat and the ball do not pass through each other when they come into contact is really the result of the electrical forces that build up between the electrons of the constituent atoms of the bat and the ball upon contact. That we cannot see through them is also related to electrons: they cause most of the incident light-rays to be reflected.

So, we know that an atom contains a tiny nucleus containing all the positive charge, with some electrons, all negatively charged, somehow circulating about the nucleus. The electrons, which carry little mass and occupy virtually no volume at all, must somehow be attached to the nucleus through the force of electrical attraction of unlike charges, much as the Earth and all the other planets are attached to the Sun through the attractive gravitational force. But here is the rub. According to classical electromagnetic theory, these electrons should spiral into the nucleus, giving off radiation over a continuous band of wavelengths (radio, infrared, visible, ultraviolet, and X-rays) as they do so. However, if this were to be the case, atoms, as we understand them, would not exist, and there would be no electrons available to participate in the formation of the chemical bonds. Without chemical reactions, there is no life. In fact, we would not even be here, let alone have the electrochemical impulses travel-ing through our neurons, to ponder these matters. As it happens, atoms *do* radiate light, but only at certain wavelengths. Furthermore, atoms can *absorb* light only at those same wavelengths. Sometimes the absorption can even cause the ejection of electrons—an effect known as the *photoelectric effect*—a phenomenon that would be hard to explain if all the electrons of atoms were stuck on the nuclei. So, somehow or other, the electrons do not spiral into the nuclei of atoms, but seem to form some sort of cloud, rather like a swarm of bees, surrounding the nucleus.

Now, in order to explain the apparent stability of atoms, and a number of other atomic-scale phenomena, a new kind of physics was invented in the 1920s. This new physics is called *quantum physics*, or *quantum mechanics*. It applies to the world on the scale of the atom—the *microscopic* world, or *microworld*. By "microscopic" we do not mean anything as large as, say, an amoeba, that could actually be seen in a laboratory microscope. Rather, *microscopic* in this context means phenomena or dimensions on the scale of the atom or smaller—the size of an atom typically being the 0.000 000 000 1 (10^{-10}) of a meter. In contrast, the large-scale world of planets, golf balls, refrigerators, and so on, is often called the *macroscopic* world, or *macroworld*. It is in this world where the laws of classical physics apply. Although we shall not enter into the details here, quantum mechanics explains a wide range of phenomena: essentially all of chemistry, the solid state and the operation of transistors, light and the operation of lasers, nuclear reactions such as the thermonuclear fusion that is the source of the energy of the stars (and of the hydrogen bomb), superconductivity, and so on.

Therefore, it seems that quantum mechanics is a rather successful theory. In spite of this, it turns out that there are elements of quantum mechanics that are quite unsettling—not in regard to the predictive power of the theory, but rather to its metaphysical foundations. In applying classical physics, or in going about our business in the everyday world, we make certain "common-sense" assumptions about reality. For example, in a criminal case being tried in a court of law, each side tries to convince the jury that the defendant did, or did not, commit the crime in question. In spite of the fact that witnesses may have lied, evidence may have been destroyed or tampered with, and so on, it is still possible to say that there *is* an objective truth, albeit perhaps unknown to the jury (or maybe to anyone for that matter) regarding the events associated with the crime. Anyone who assumes that there is such an objective truth in this sense is classified as a *realist*, his point of view being *realistic*. Our use of these terms is limited; we are not comparing the all too human traits of believing in some fantasy world as opposed to a realistic world outlook. Here, a realist asserts that, to borrow a line from a popular television series, "the truth is out there", independent of anyone's mind. We call that truth *objective reality*. In classical physics, and in all of science, the assumption of objective reality is implicit. Perhaps, about some distant star, there may, or may not, be an orbiting planet. Because of the great distance, we may not know of the planet's existence, but we can say that if it does exist, it will be obeying Newton's laws of motion and the law of gravitation. In other words, the objective existence and behavior of the planet are independent of whether or not anyone on Earth *knows* about it. Closer to home, to pick a rather overworked example, consider the question: When a tree falls in the forest, does it make a sound if no one is there to hear it? Well, if what we mean by "sound" is just the mechanical vibration of the air which can be detected by a tape recorder, even if no one is there, the "common-sense", objectively realistic, answer to this question is "yes". Of course, if what we mean by "sound" is the sensation produced in the ear and the brain, then, of course, if no one is there, there is no sound.

When we speak of "common sense" we generally mean certain rules of logic known as *Boolean logic*—named after George Boole, who summarized his studies on logic in a book entitled *An Investigation into the Laws of Thought*. These rules appear to be universal in the sense that anyone correctly using them would arrive at the same conclusions under identical circumstances. Electronic digital computers rely on the rules of Boolean logic to carry out their functions. But the microscopic world described by quantum mechanics does not always seem to work accordingly. As this is not easy to grasp—it is not, after all, "common sense"—perhaps it is best to proceed with some examples in order to exhibit the sharp contrast between the behavior we expect in the macroscopic world and what actually happens in the microworld. In reality, even in the everyday macroworld, "common sense", at least as far as science is concerned, is not as powerful an idea as it is popularly made out to be. For example, Newton's first law of motion says that an object moving at constant velocity and experiencing no outside forces should remain in the same state of motion. As simple as the law appears to be, it is not common sense. Early scientists used to think that to keep something moving one had to *constantly* apply a force and that the object would come to rest upon cessation of the force. Casual observations certainly seem to confirm as much. If one kicks a tin can, it will eventually come to rest. What brings it to rest, though, is the friction force between the asphalt and the can, and it is this not-so-*obvious* force that violates the conditions for the first law to be valid. Physics, and science in general, is filled with all sorts of cases where "common sense" fails. But in quantum mechanics there are instances where classical notions, confirmed as valid in the everyday world to which classical physics applies, common sense or otherwise, seem to break down. These breakdowns usually occur in the microscopic realm, which is far removed from our normal experience, but there are a few cases, to be examined in a later chapter, where breakdown occurs on a larger scale.

2.2 "Quantum" Coins

We consider first a simple experiment where a coin is tossed onto a flat level surface. There are two possible outcomes for each run of the experiment: *heads* or *tails*. Consider the following sentence: If a coin is on the surface and shows either heads or tails, then it follows that the coin is either on the surface and shows heads, or is on the surface and shows tails. The logic of this statement seems inescapable, though the conclusion seems rather trivial. It is easy to construct logical statements of this type, because we know how to use the connective words *and* and *or*. Furthermore, it is equally obvious, and in fact would normally go without saying, that the act of *observing* the coin has no effect on whether or not it shows heads or tails—the state of the coin being objectively *definite*.

In contrast, if the coin were a quantum mechanical one, the simple logical statements above might not apply. In the quantum world we would be able to introduce two possible quantum states (also known as state vectors) that would describe the possible outcomes of measurements on the state of the coin. In the case where the coin is heads we could denote the state as $|h\rangle$, and if tails, $|t\rangle$. The angular bracket $|\rangle$ is known as a *ket* (or a ket-vector) where the label inside it—in this case the h or the t—specifies the state of the quantum system.[†] A ket

---

[†] It was the physicist P. A. M. Dirac, one of the founding fathers of quantum mechanics, who introduced, in the late 1920s, the kets, or ket vectors, $|\rangle$. He also introduced what he called "bra" vectors, denoted $\langle|$, though we shall avoid using them in this book. The names "bra" and "ket" come from the notion of taking a *bracket* in the form $\langle\ \rangle$ as used to indicate averages, and splitting into $\langle|\ldots|\rangle$, or bra-ket. Dirac was, to say the least, a very strange fellow, whose personality almost certainly fell somewhere within the spectrum of Asperger's syndrome/autism. It seems that he was unaware at the time that the word "bra" had a commonly understood meaning. See the biography of Dirac by G. Farmelo, listed in the bibliography for this chapter.

is *not* a quantity that carries a numerical value. The states $|h\rangle$ and $|t\rangle$ represent the possible outcomes of a coin-toss, and are examples of states connected to outcomes of the measurement of some physical property of a system—in this case the property that the coin displays heads or tails. These are just the same outcomes that are possible for a classical coin. If it is known to be heads its state is $|h\rangle$, and if it is known to be tails its state is $|t\rangle$. The label inside the ket notation is a statement of what is known about the coin. On the other hand, a true "quantum" coin *could* be in a *superposition* state, denoted $|S\rangle$, of the two possible states. An example of such a state is

$$|S\rangle = \frac{1}{\sqrt{2}}|h\rangle + \frac{1}{\sqrt{2}}|t\rangle = \frac{1}{\sqrt{2}}(|h\rangle + |t\rangle).$$

In this case, the label S inside the ket indicates what is known about the state of the coin—that in this case it is a *superposition* of the states heads and tails. The addition sign here does not imply any kind of arithmetic operation, nor does it imply that the state of the coin is simultaneously heads and tails, as we shall discuss shortly. The meaning of the addition sign will soon become clear. The numbers $1/\sqrt{2}$ multiplying each of the kets are known as probability *amplitudes* associated with the possible outcomes of a measurement. The square of a probability amplitude yields the probability of obtaining the corresponding outcome of a measurement. In this particular state, the probability of obtaining h is $\frac{1}{2}$, and that for obtaining t is also $\frac{1}{2}$. Note that these probabilities add up to 1. Other possible superpositions can be considered, though they all have the form

$$|S\rangle = a_h|h\rangle + a_t|t\rangle,$$

where the sum of the squares of amplitudes a_h and a_t must add up to 1: $a_h^2 + a_t^2 = 1$, meaning that the sum of the probabilities for the occurrence of each of the possible outcomes must add to 1.[‡]

[‡] In actuality, probability amplitudes are complex numbers. A complex number a has the form $a = x + \iota y$ where $\iota = \sqrt{-1}$. If a is a probability amplitude, then the associated probability is given as $|a|^2 = x^2 + y^2$, a number that would have to be less than, or equal to, 1 for it to be a probability. So if a_h and a_t are complex numbers, then one must replace the condition $a_h^2 + a_t^2 = 1$ with the condition $|a_h|^2 + |a_t|^2 = 1$. In general we shall ignore this complication in dealing with probability amplitudes.

We have been calling things denoted by the symbol | ⟩ *ket vectors*, but we have not said why. The word *vector* connotes a quantity that has both a magnitude and a direction. *Velocity* is an example of a vector; it is specified by a magnitude (how fast something moves, or its *speed*) and its direction. A velocity of 60 mph north is different from a velocity of 60 mph east. The speeds are the same but the directions are different. On the other hand, if the velocity is specified as 60 mph at 30° north of east, as in Figure 2.1, then we say that along the east–west direction the velocity is 60 mph × cos30° = 52 mph (east), and along the north–south direction it is 60 mph × sin30° = 30 mph (north). We call these numbers (52 mph, 30 mph) the *components* of this velocity vector. Note that because of the Pythagorean theorem (the sum of the squares of the sides of a right triangle equals the square of the hypotenuse) we must have $(52 \text{ mph})^2 + (30 \text{ mph})^2 = (60 \text{ mph})^2$.

We can abstract from this the notion of a *vector space*, which is not necessarily tied to the directions of ordinary physical space, but can be thought of entirely as a mathematical construct. The space can have as many dimensions as is needed to accommodate all the attributes necessary to describe the system in question. For the quantum coin

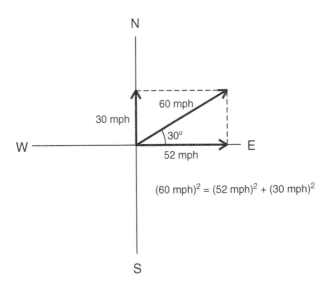

Figure 2.1 A velocity vector of magnitude 60 mph directed at 30° north of east has the components 52 mph east and 30 mph north.

we have a two-dimensional vector space corresponding to the ket vectors $|h\rangle$ and $|t\rangle$ —the possible outcomes for a heads–tails experiment. One can think of these ket vectors as though they specify two perpendicular "directions" in this abstract space, analogous to north–south and east–west in the example of the velocity. The probability amplitudes a_h and a_t are understood to be the *components* of the vector $|S\rangle$, in the same sense as the numbers 52 mph and 30 mph represent the components of the velocity vector discussed above. We illustrate this idea in Figure 2.2. Here, though, the numbers a_h and a_t are constrained by the condition that $a_h^2 + a_t^2 = 1$ —again because the probability of obtaining heads is a_h^2, while that of obtaining tails is a_t^2 — and as these are the only possible outcomes for a heads–tails experiment, these probabilities must add up to 1.

The states $|h\rangle$ and $|t\rangle$ have clear meanings as outcomes of measurements on the state of a coin, even for a quantum coin. But what can be said about a quantum coin in a superposition state $|S\rangle$? What does the superposition state *mean*? There is no easy explanation, because a real coin is a classical object and can never be in a superposition state. We ask the reader, at least for the moment, to suspend disbelief. Shortly, we will discuss a real quantum phenomenon that we hope will convince you that our interpretation of superposition does make some sense. Let us first talk about a real, classically-behaving coin. We flip it and obtain either heads or tails, each with probability $\frac{1}{2}$, assuming that the coin is not loaded in some way. The

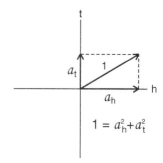

$$1 = a_h^2 + a_t^2$$

Figure 2.2 The probability amplitudes a_h and a_t of the state vector $a_h|h\rangle + a_t|t\rangle$ form the components of a vector with the restriction that the sum of the squares of the amplitudes must add up to 1.

state of a real coin, in the language of quantum mechanics, is what is called a *mixed state*, or a statistical mixture (denoted $_{\mathrm{Mixed}}$), which we represent by the notation

$$\rho_{\mathrm{Mixed}} = \begin{cases} \frac{1}{2} \text{ for } |\mathrm{h}\rangle \\ \frac{1}{2} \text{ for } |\mathrm{t}\rangle \end{cases}.$$

For our purposes, this notation means that the probability of obtaining h or t is $\frac{1}{2}$. We shall use this symbol to mean that the state of the coin, after flipping but *before* measurement, is h *or* t with the corresponding probabilities. There are no probability amplitudes here, only probabilities. And for this reason there is nothing inherently quantum-mechanical about the mixed state even though it does have a probabilistic nature. We can assert that the state of the coin is objectively definite (h or t), even if we know only the associated probabilities ($\frac{1}{2}$ for h or t in this case).

In contrast—and again we beg the reader's indulgence on this point—a coin in the superposition state $|\mathrm{S}\rangle$ has no definite state with respect to heads or tails. Rather, the attribute associated with these possible outcomes is objectively *indefinite* (or indeterminate) until, that is (and this where we might seem to be throwing common sense out the window), someone, or something, actually performs a measurement to find out (that is, looks to see). It is not merely that we do not know what the state of the coin is with respect to heads and tails, but rather that it has no well-defined attribute of heads or tails at all. Its state *is* the superposition of the heads and tails states. Thus, with respect to those possibilities, the coin is in a state of limbo. There is no objective reality to the state of the coin other than to say that it is in a superposition. In fact, the only definite thing we can say *is* that it is the superposition state! In a very real sense, ordinary language fails us in attempting to describe some attributes of particles in quantum-mechanical superpositions.

Surely, the idea that an object can be in a superposition of possible states with respect to some observable quantity does violence to our "common sense". Furthermore, according to the usual interpretation of quantum mechanics, the act of observation "collapses" this indefiniteness onto one of the definite states: heads or tails. The indefiniteness contained in a superposition, and the notion of the collapse of the state vector, are the essence of the so-called *Copenhagen interpretation* of quantum mechanics—so named because of its close association with Niels Bohr and his school of physics in that city.

Throughout this book we will have much to say about this interpretation. But if all the preceding seems nonsensical, that might be because a real coin is not described by the laws of quantum mechanics. In fact, it is not at all obvious, from the discussion above, as to why anyone would ever have to think in such a convoluted way. It is hardly possible to imagine how there could even be observable consequences for a coin in a superposition of the states heads and tails. We could imagine that the surface the tossed coin lands on itself is a kind of measuring device that interacts with a tumbling coin. When the coin comes to rest, it must have one of the states: heads or tails. For a real coin we expect that it has one of the sides facing up just *before* coming to rest on the surface. But according to the Copenhagen interpretation of quantum mechanics, the quantum coin would not have a definite side facing up *until* it comes to rest on the surface, and the subtle distinction between these two scenarios is of utmost importance.

To fully appreciate the necessity for something as seemingly obscure as the Copenhagen interpretation of quantum mechanics, we have to turn over the discussion to the behavior of a truly quantum-mechanical system, and show how actual laboratory experiments force this strange picture of the microworld upon us. It is important to understand that the Copenhagen interpretation did not arise in a vacuum, so to speak, nor is it some kind of philosophic posturing. Rather, it arose as the result of the confrontation with phenomena not readily interpreted in terms of our "common-sense" concepts of the world. Thus, it is a pragmatic attempt to describe the nature of the microworld, as revealed in the type of experiment we are about to describe, in language that makes sense to us. And ultimately, as you will see, it does not really make "sense" (or at least "common sense") from a classical point of view, even though the Copenhagen interpretation is internally consistent.

Incidentally, the basis of current digital technology—the digital computer, for example—is the ability of conventional electronic devices to store and manipulate information in discrete units called *bits* (a term compressed from *binary digits*), where a bit can be one of two distinct values, usually denoted by 0 and 1. Physically, the bits represent different voltage levels. But we can also use distinct quantum states of some system to represent the bits (they are actually called *qubits*) to make possible a new kind of computing technology: quantum computing. Quantum computers (if they can be built), because of the ability to create superpositions of the qubits, allow for computations to be performed in a highly parallel, and efficient, manner, which no classical computer can match. We will say more about this in Chapter 6.

2.3 Superpositions versus Mixtures

The reader may be wondering about the practical difference between the superposition state

$$|S\rangle = \frac{1}{\sqrt{2}}(|h\rangle + |t\rangle),$$

and the mixed state

$$\rho_{\text{Mixed}} = \begin{cases} \frac{1}{2} \text{ for } |h\rangle \\ \frac{1}{2} \text{ for } |t\rangle \end{cases}.$$

After all, according to what we discussed above, the measurement of the state of the coin gives either heads of tails with probabilities $\frac{1}{2}$ for each outcome, whether it is in the superposition or the mixture. The probabilities can be verified by repeated measurements on an identically prepared coin. After many runs of the experiment are performed—say 1,000—about 500 of the runs produce heads and, of course, the balance will be tails. The probabilities are defined, in practice, as the ratio of the number of outcomes with heads or tails to the total number of runs of the experiment. The more runs of the experiment, the closer the ratios approach the predicted probabilities. So, how can we tell the difference between superpositions and mixtures?

A partial answer to this question is that quantum superposition states can exhibit the phenomenon of *interference*, whereas mixtures cannot. So, now we need to discuss the phenomenon of interference. In the context of classical physics where interference is intrinsically a phenomenon associated with waves and not with particles. But in the quantum world, even things known to be particles can apparently give rise to interference effects. In the next section we discuss classical wave phenomena and introduce the notion of interference.

2.4 Light, Waves, and Interference

We shall begin with a classical picture of light. According to Samuel Johnson, "We all *know* what light is, but it is not so easy to *tell* what it is." Here we shall not review the history of our understanding of light, but it is worth pointing out that Newton thought that light is a stream of particles, or *corpuscles*, as he called them. On the other hand, Christiaan Huygens, among others, thought that light is some kind of wave phenomenon, due to its observed ability to propagate around the edges of a barrier. This process, called *diffraction*, is a phenomenon that particles should not exhibit. The wave nature of light was convincingly demonstrated by Thomas Young in 1802. He showed that light could be made to exhibit *interference*—a phenomenon known to be associated with waves. As interference will be of great importance in what follows, we illustrate the origin of the effect using a simple graphical representation of waves. We shall take our waves as *transverse* sinusoidal waves, which have the shape shown in Figure 2.3. *Transverse* means that the oscillations are perpendicular to the direction of propagation. The wave in Figure 2.3 is actually just a section of a wave "train", where the wave really should be thought of as moving from left to right, as indicated by the arrow.

We now introduce some terminology. The distance between successive wave-crests along the direction of propagation is called the *wavelength* (represented by the symbol λ), and the number of crests

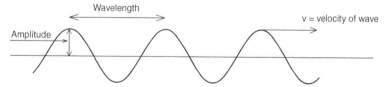

Figure 2.3 Profile of a wave moving at the speed of light, c. The wavelength λ is the distance between successive crests, while the frequency f (not shown) is the number of crests passing a point in 1 second. The amplitude of the wave is half the distance from crest to trough.

passing by any point in 1 second is the *frequency* (represented by the symbol f). For a light-wave, the speed of the wave-crests passing an observer is the speed of light, denoted c, which happens to be about 300,000 km/sec. The speed, wavelength, and frequency of light are related by the simple formula $\lambda \times f = c$ (or simply $\lambda f = c$). The wavelengths of visible light range from about 400 to 700 nm. We, through our eyes and brains, perceive the wavelengths of light as colors. Light of wavelength 400 nm is perceived as violet, while light of 700 nm is perceived as red. All the other colors correspond to different wavelengths—one color gradually blending into another with changing wavelength. Of course, were we referring to a wave on the surface of water, the velocity would be much, much less, and the wavelengths much, much longer. Note that there is an inverse relationship between wavelength and frequency as long as the wave velocity is the same: the greater the wavelength the smaller the frequency, and vice versa. The amplitude, or height, of the wave measured from the "zero" point of the oscillations is the greatest amount of displacement, perpendicular to the direction of propagation of the wave, of whatever is waving (such as water). Now, when two waves come together the perpendicular displacements at all points where the waves overlap simply add or subtract to produce a resultant wave which is the superposition of the original two waves. Just what this new wave looks like depends on the *phase* relation between the original two waves. The phase relation refers to the degree to which the crests and troughs of the two waves overlap each other. When they overlap exactly, the waves are said to be perfectly *in phase*, and the resultant wave has an amplitude equal to the sum of the amplitudes of the original waves, as illustrated for two waves of equal amplitude in Figure 2.4. This is called *constructive interference*. On the other hand, if the waves come together in such a way that the crests of one line up with the troughs of the other, the waves are said to be completely *out of phase*. The result is a cancellation when the crest and troughs are added together, since the amplitudes are in opposite directions. We illustrate this in Figure 2.5—again for the case when the original waves have equal amplitudes. This is *destructive interference*. If the waves are neither exactly in nor out of phase, there is a *partial* cancellation, as shown in Figure 2.6. Interference effects can be seen easily in water waves if two pebbles are simultaneously dropped some distance apart into a pool. Circular

Crests and troughs align

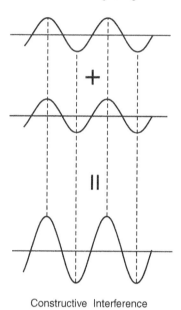

Constructive Interference

Figure 2.4 When two waves come together in phase, meaning that the crests and troughs align, the amplitudes add to produce constructive interference.

waves will spread out from each drop point, as illustrated in Figure 2.7, where each solid-line circle represents a wave crest, and each dashed-line circle a trough. Where crests overlap we have constructive interference, and where a crest overlaps with a trough we have destructive interference.

If light is a wave, then one might reasonably ask: *What is waving?* It turns out that light-waves are actually composed of oscillating electric and magnetic fields propagating freely though space—even empty space. Electric fields originate from particles carrying electric charges, while magnetic fields originate from the *motion* of electric charges (such as electric currents). The fields fill the space around the charges, and currents and can be sensed by their effects on the motions of other charged particles placed in the fields. Rapidly oscillating charges can produce electromagnetic waves in which the electric and magnetic fields oscillate sinusoidally at right angles to each other, as shown in

Crests and troughs out of phase

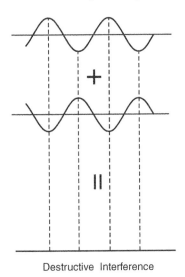

Destructive Interference

Figure 2.5 When two waves come together out of phase such that the crest aligns with troughs, the amplitudes cancel to produce destructive interference.

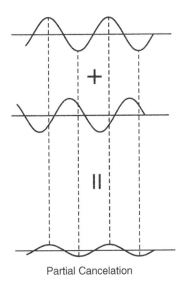

Partial Cancelation

Figure 2.6 If two waves come together neither completely in nor out of phase, there is a partial cancellation of amplitudes.

(a) Circular wavefronts   (b) Interference

(c)

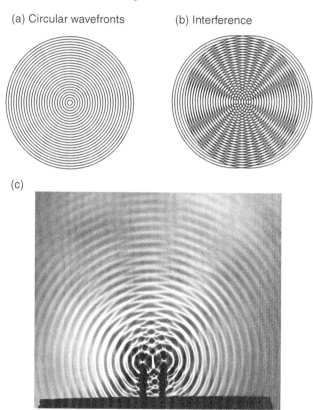

Figure 2.7 (a) This shows a set of circular wave-fronts generated by dropping a pebble onto a calm water surface. (b) If two such pebbles are dropped simultaneously nearby, the resultant circular waves from each set up an interference pattern. (c) A photograph of interference between two sets of circular water-waves.

Source: Photo Researchers.

Figure 2.8. Our eyes—or rather, the photosensitive cells in our eyes—detect the electric fields of light-waves in the wavelength range 400–700 nm—the visible part of the electromagnetic spectrum. In fact, the spectrum of electromagnetic waves is broad: gamma-rays (emitted by atomic nuclei), X-rays, ultraviolet rays, visible light, infrared rays, and radio waves are fundamentally all the same. The only differences among these waves are their wavelengths and frequencies (the list

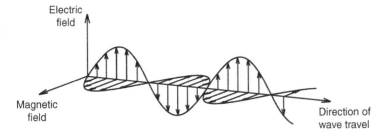

Figure 2.8 A light-wave consists of electric and magnetic fields oscillating at right angles to each other and perpendicular to the direction of propagation. The orientation of the electric oscillations defines its direction of polarization.

just given is in order of increasing wavelength and decreasing frequency—the higher the frequency, the higher the energy carried by the waves).

The direction along which the electric field of a light-wave oscillates is called the *direction of polarization*, or just *polarization*. Light from ordinary sources—from the Sun, incandescent light bulbs, and so on—is *unpolarized*; that is, the electric fields are oscillating in random directions, as illustrated in Figure 2.9(a). By using polarizing filters (made of the material used in polarizing sunglasses) one can filter out all the light from a beam except for that of a particular polarization direction, thus producing a polarized beam of light. The direction of the polarization depends entirely upon the orientation of the filter, as shown in Figure 2.9(b), where the light is polarized at an angle θ to the horizontal. For future reference, we illustrate in Figures 2.9(c)–(e) the specific choices of the angle θ —$0°$ for horizontal (H), $90°$ for vertical (V), and $\pm 45°$, respectively. Incidentally, most lasers also produce light that is polarized. In the discussion that follows we shall assume that the light is polarized, though the direction of the polarization is generally irrelevant.

Because the wavelength of light is so much shorter than that of water-waves, a little more finesse is required to observe interference with it. The experiment of Young goes like this: A light beam is passed through a slit in a barrier and then encounters a second barrier having two slits, both slits being the same distance from the slit on the first barrier. The beams from these slits then fall onto a screen, creating alternating bright and dark bands. A sketch of this experiment is given in Figure 2.10. These bands are the result of interference, and are

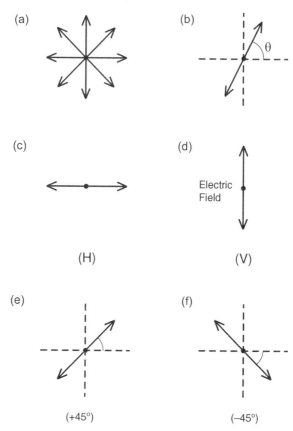

Figure 2.9 The polarization of light as seen end-on as if the light is emerging from the page, propagating to the viewer. (a) Representation of a collection of light-waves emerging from the page but having different—in fact random—polarizations. (b) A light-wave polarized at an angle θ with respect to the horizontal. (c) A light-wave with a horizontal (H) polarization. (d) A light-wave with a vertical (V) polarization. (e) and (f) Light-waves polarized at $+45°$ and $-45°$ from the horizontal.

called *interference fringes*. The explanation of the appearance of these fringes is as follows: The light-beams emerging from the two slits on the second barrier are *in phase* with each other because they both originate from a common source the same distance from each slit. But the light from each of the slits spreads out and propagates in all directions, just as in the case of water-waves from the simultaneously

a)

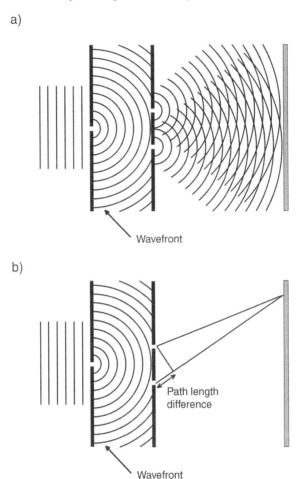

b)

Figure 2.10 Young's experiment. (a) Two pinholes or slits produce, from a single source of light, a set of waves that give rise to interference fringes much like the pattern seen in (b) The nature of the interference fringes at any point on the screen depends on the difference of the difference of the lengths of the paths from the point and the two pinholes.

dropped stones, apart from the presence of the screen with the slits. Thus each point on the screen receives light from each of the slits, but the distances traveled by each wave will be different. That being the case, at some points on the screen the light waves from the slits will be in phase, giving rise to constructive interference resulting in the bright fringes. At other points on the screen the light-waves will be out of phase, and destructive interference will occur, giving rise to the dark fringes. At points where the light-waves are neither exactly in phase or out of phase, there is only partial interference, so that the bright and dark fringes gradually blend into each other. Figure 2.11 is a photograph of the interference fringes obtained in Young's experiment, but with a laser as the source of light.

Incidentally, it is possible for the reader to see interference fringes for himself or herself. Shine a laser pointer onto a screen and place a hair in the laser beam. Interference fringes arise on the screen as the result of light going both ways about the hair. *Warning: never look directly into a laser beam!* Alternatively, interference can be seen even more directly by looking at a distant streetlight with a hair in front of the eyes.

Perhaps we should emphasize yet again that everything involved in the explanation of this interference is classical—interference being a

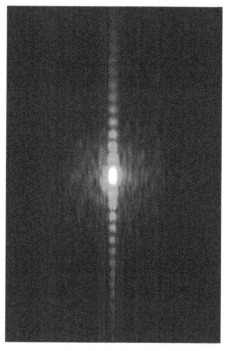

Figure 2.11 A photograph of a Young's double-slit interference pattern, created using a laser beam. (Photograph courtesy Saaber Shoyeb.)

characteristic of all wave phenomena. In sharp contrast to waves, which are obviously extended over large regions, particles are localized objects, and we do not expect to find interference effects in experiments involving them. For the sake of highlighting this contrast, we now consider essentially the same experimental set-up as above, but where particles are sent, one at a time, through the double slits.

To make this as dramatic as possible, we will take our particles to be bullets and the source a machine gun. Imagine that we have a steel plate that has two narrow slits cut through it wide enough so that the bullets can pass though. The machine gun will produce a spray of bullets over the plate. We imagine that at some distance behind the plate a thick wooden "screen" is placed to catch the bullets that pass through the slits. This arrangement is pictured in Figure 2.12. Now, suppose we close off one of the slits by bolting a piece of steel plating over it. This ensures that the bullets that collect on the screen come only from the remaining open slit. If we fire the gun many times we will find a clustering of bullets embedded in the wooden screen just opposite the open slit, as shown on screen (a). Obviously, if we now close this slit and open the other, the same thing will happen on that side, as shown on screen (b). When both slits are open we just get two overlapping clusters of bullets, as shown on screen (c), and evidently we do not get anything at all that looks like the interference fringes in the experiment with the light-waves. By the way, nothing about this experiment would change if some other kind of gun—one that fires at a slower rate—were used instead of a machine gun. All the machine gun does is cause rapid accumulation of bullets on the screen. "Common sense" dictates that the lack of interference fringes is directly related to the fact that bullets are particles, not waves, and the fact that they pass through the slits one at a time. Particles do not give rise to interference effects.

Or do they? The idea that material microscopic particles such as electrons might have wave-like properties predates any experimental evidence. In 1924 the French physicist Louis de Broglie, struck by the fact that light has both wave-like and particle-like properties (the particle-like nature of light—the particles being *photons*—is discussed in the following chapter), suggested that nature might be symmetrical, and that material particles might have wave-light properties as well. He postulated that the wavelength associated with particles be given by the simple relationship $\lambda = h/p$, where h is Planck's constant (a very small number) and p is the momentum of the particle. The

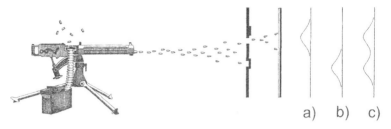

Figure 2.12 Young's experiment performed with bullets. If the lower hole in the iron plate is closed off, all the bullets that make it through cluster opposite the open hole as indicated on screen (a). If the upper hole is closed off and the lower one opened, the bullets cluster opposite the lower hole as indicated on screen (b). If both holes are open, bullets cluster opposite both holes as indicated on screen (c), and there is no interference pattern.

momentum of a particle is its mass (m) times its velocity (v): $p = mv$. In 1925, two American physicists, C. Davisson and L. H. Germer, working at Bell Telephone Laboratories, discovered the wave nature of electrons in an experiment where a beam of electrons bombarded a crystal of nickel atoms. Interference fringes were seen in the pattern of the scattered electrons, and the associated wavelength was found to be in good agreement with the prediction of de Broglie.

In the Davisson–Germer experiment the electron beam impinging on the crystal contained many electrons simultaneously. But what would happen if we could carry out an interference experiment with electrons one at a time, very much like the one-bullet-at-a-time experiment described above? Would we still get interference?

2.5 Interference with Electrons

When you come to a fork in the road, take it.

YOGI BERRA

Bullets are certainly classical objects. They are macroscopic, and obey the well-established laws of classical mechanics. Let us consider electrons, which are also known to be particles. Because electrons are charged particles, their motion can be manipulated by electric and magnetic fields. This is what happens inside old-fashioned television tubes and computer monitors (though not the newer flat screens) in order to form images. Electrons can be detected indirectly by observing the ionization tracks they leave when passing through a cloud chamber, as shown in Figure 2.13. These tracks are just lines, and seem to be in no way related to waves—more evidence of the particle nature of electrons. The electron is an elementary particle. It seems to have no substructure (that is, it is not a composite of other particles) and, as mentioned in Chapter 1, it seems to have no extension to it at all: it appears to have the dimensions of a mathematical point. It is, in a sense, the very embodiment of what it means to be a particle. Furthermore, in the most accurate physical theory that we have, quantum electrodynamics (QED), calculations are done assuming the electron to be a point particle. The calculations agree with experimental results to about 1 part per billion, which is fairly convincing evidence that the electron is very close to being a point particle.

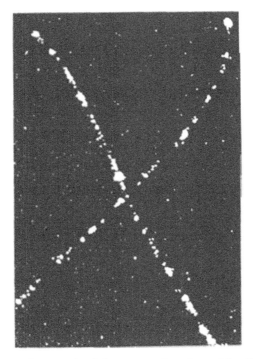

Figure 2.13 A photograph of electron tracks, taken with a Wilson cloud chamber.

2.6 Interference with Electrons, One at a Time

Now, suppose we carry out the double-slit experiment once more, this time using electrons. The wave nature of electrons has been observed in various experiments since the 1920s. However, surprisingly, an experiment analogous to the one with bullets described above—that is, with one electron at a time, though much discussed in the earlier literature as a "thought" experiment—was achieved for the first time only in 1989, by Akira Tonomura and collaborators at the Hitachi Advanced Research Laboratory in Japan. Actually, these experimenters did not use a barrier involving parallel slits, but rather an electron microscope equipped with a device called an *electron biprism*, which is equivalent to a set of double slits. This device, pictured in Figure 2.14, consisted of a set of parallel plates with a thin wire between them. The wire had a net positive charge compared to the two plates, which are grounded. This caused electrons to be deflected, as pictured, when they passed through the apparatus. The electrons then fell onto a fluorescent film which emitted pulses of light—one pulse for each electron. The light-pulses were subsequently detected by a special TV camera so that the location of each electron's hit on the film could be displayed on a monitor.

The experiment was performed with the source of the electrons turned down so low that there would be only one electron at a time between the source and the film. This is essentially the same condition for the gun, which fires only one bullet at a time. Therefore, you might think that the pattern of the electrons on the screen would be just like that for the bullets in the case with both slits open. However, you would be wrong. Fig. 2.15 shows a sequence of photographs taken at different times, with an increasing number of electrons collected on the film. In the figure, photograph (1) has only ten electrons, (2) 100 has electrons, (3) has 20,000, and (4) has 70,000. With only ten electrons there is no discernible pattern in the dots. The same can be said after 100 electrons have been collected. But after enough electrons have struck the film, an interference pattern becomes visible, as is evident when 20,000 electrons have been collected, becoming even more pronounced after 70,000 have struck the film.

How can there be an interference pattern when there is only a stream of particles passing through the apparatus one at a time? Clearly, these particles are not behaving like bullets. Furthermore, the pattern cannot be due to interactions among the electrons—at

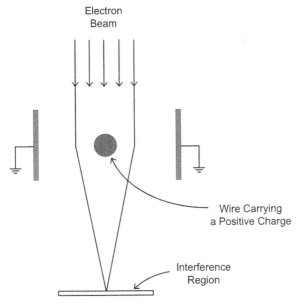

Figure 2.14 A schematic of the apparatus used by Tonomura to obtain interference with electrons one at a time. The device is called an electron biprism. The parallel plates are electrically grounded and the wire in the center carries a positive charge. The resultant electrical forces on the electrons focus them onto the interference region as indicated.

least not directly, since at any given time there is at most only one electron between the source and the film. The conundrum is this: How can there be an interference pattern associated with the *collective* behavior of the electrons when the electrons pass one at a time through the apparatus and thus do not seem to be able to influence each other? This is really quite spooky. Could it be that each electron somehow goes around *both* sides of the wire simultaneously and interferes with itself? Well, at the very least we may deduce from this that though electrons may be point particles, they do not behave like *classical* point particles. Evidently, in spite of the fact that the electrons *are* particles (note that each electron is "seen" as just a point on the film) there is something wave-like in their behavior. A duality exists between their wave and particle natures. Nothing like this has

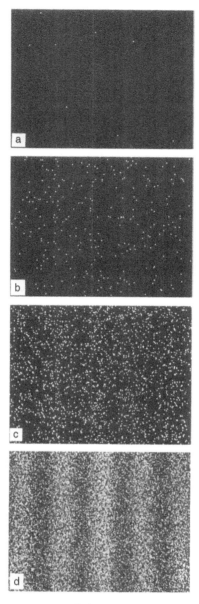

Figure 2.15 A time sequence of photographs with an increasing number of electrons, each arriving one at a time, collecting on the plate. Photograph (a) has only ten electrons, (b) has 100 electrons, (c) has 20,000, and (d) has 70,000. After a sufficient number of electrons have been collected, an interference pattern emerges, even though the electrons are spaced far enough in time so that only one at a time passes though the effective double-slit arrangement of Figure 2.14. Used with permisson from the American Institute of Physics.

ever been seen in the behavior of macroscopic particles such as bullets. What is happening here?

The fact that the particles (the electrons) are acting in some sort of wave-like manner naturally leads us to introduce the notion of an associated quantum *wave function*, which we denote Ψ. Historically, the notion of the wave function came about shortly after de Broglie proposed the wave nature of particles. In 1925 the Austrian physicist, Erwin Schrödinger, reasoned that if particles can act like waves then there must be a wave equation to describe those waves (the equation he proposed is now called the Schrödinger equation), and that associated with this equation there had to be wave functions. The value of the wave function depends on position in space and on time. If the particle in question is well-localized in some region of space, the corresponding wave function will have the value of zero outside of that region. The quantum wave function does not have direct physical meaning in contrast to, say, the direct physical meaning of an oscillating electric field of a light wave. Electric fields are real things that can exert real forces on charged particles. Quantum wave functions are not physical in that sense: they exert no forces. Rather, a quantum wave function is related to probability—the probability of finding the particle in question in a region of space. Strictly speaking, in general, a quantum wave function can be a *complex* function, meaning that it could contain a part that is multiplied by $i = \sqrt{-1}$. That is, a wave function Ψ has the form $\Psi = \Psi_R + i\Psi_I$, where Ψ_R and Ψ_I are called the real and imaginary parts of the function, Ψ_R and Ψ_I themselves being *real* functions (that is, not containing $i = \sqrt{-1}$). We can define the *complex conjugate* of Ψ written as Ψ^*, which is what we obtain by replacing i by $-i$ in Ψ to obtain $\Psi^* = \Psi_R - i\Psi_I$. The probability of finding the particle is proportional to the quantity $\Psi^*\Psi$, which is usually written as $|\Psi|^2$, and is equivalent to $\Psi_R^2 + \Psi_I^2$. The quantity $|\Psi|^2$, often referred to as the "square" of the wave function, is essentially a probability. The wave function itself is proportional to what is called a *probability amplitude*, whose square is a probability.[§] Wave functions are just one type of probability *amplitude* that we shall encounter (not all probability

---

[§] Technically, $|\Psi|^2$ is a probability density, or probability per unit volume of space, for finding the particle, but we shall not need to be concerned about these distinctions.

amplitudes are wave functions). A rule in quantum physics is that *probabilities* are obtained by *squaring probability amplitudes*. Another rule states that we add probability amplitudes associated with all the different ways one can go from an initial state to a final state. We apply that rule immediately to explain the interference pattern of the electrons on the screen in the Tonomura experiment.

The explanation provided by quantum mechanics is as follows: The wave function of an electron on the film, Ψ, is a superposition of the wave functions Ψ_1 and Ψ_2 associated with the two possible paths to reach the film through the two double slits. Assuming each path is equally probable, the wave function of the electron just before striking the screen is the superposition

$$\Psi = \Psi_1 + \Psi_2.$$

To be clear, referring to Figure 2.16, Ψ_1 is the wave function associated with an electron passing through slit 1, and Ψ_2 is associated with an electron passing through slit 2. The location of an electron at any given time before striking the screen is indeterminate in the following sense: The electron has no definite position; only a certain probability that it could be found in any particular place if a measurement is performed to find out. Quantum mechanics tells us how to determine the wave function (or the state vector).

Now, if either slit is closed off, no superposition is possible as there is only one way for electrons to go. Thus, the wave function on the film will be given by either Ψ_1 *or* Ψ_2, depending on which slit is open, and the probabilities are just the squares of these numbers, $|\Psi_1|^2$ and $|\Psi_2|^2$. These will look very much like the distribution of the bullets in the case when only one slit is open. However, when both slits are open the wave function is given by the Ψ above, from which we obtain the probability

$$|\Psi|^2 = |\Psi_1 + \Psi_2|^2 = \left(|\Psi_1|^2 + |\Psi_2|^2 + \Psi_1^* \Psi_2 + \Psi_1 \Psi_2^* \right).$$

The terms $|\Psi_1|^2$ and $|\Psi_2|^2$ are just those corresponding to each of the two slits, as we just mentioned. But the terms $\Psi_1^* \Psi_2 + \Psi_1 \Psi_2^*$ are new and would not be present were we only adding probabilities and not probability amplitudes. It is these last terms, which can take on negative as well as positive values, that are responsible for the interference fringes. The wave functions Ψ_1 and Ψ_2 corresponding to

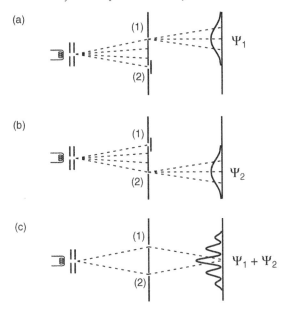

Figure 2.16 The quantum-mechanical explanation of the double-slit experiment with electrons. (a) and (b): Ψ_1 and Ψ_2 are the quantum amplitudes associated with the electrons passing through slits 1 and 2 respectively, and produce probability distributions, proportional to the squares of these amplitudes, similar to those for the case for the bullets, as in Figure 2.12. (c) With both slits open the amplitudes from both paths add, and because the amplitudes can take on both positive or negative values, and because they can take on values opposite to each other, interference fringes in the probability distribution are present.

the paths taken through the double slits to reach the same point on the screen will each possess different phases depending on those path lengths. That is, they will oscillate with respect to location but with different phases, just as in the case for light-waves, as pictured in Figures 2.4–2.6. If the phases are such that Ψ_1 and Ψ_2 are both positive or both negative, then $|\Psi_1 + \Psi_2|^2$ will be large and we will get constructive interference corresponding to a bright fringe where an electron is highly likely to be found—these places being the sites of the clustering of the electrons on the film. On the other hand, when they are of opposite signs, destructive interference can occur to produce dark fringes where the probability of finding an electrons is low—that is, the "dark" bands across the film where there are relatively few electron hits. For example, if $\Psi_2 = -\Psi_1$ at some point then the total wave function at that point must be zero, so there is zero probability that an electron will be found there. The central point of all of this is that *electrons do not interfere with each other: it is the probability amplitudes that interfere*. It is sometimes said that "each electron interferes with itself", although this should not be taken literally. Ultimately, it is the probability amplitudes for a single electron that interfere. That wave functions, and quantum amplitudes generally, are related to probabilities was first suggested by the German physicist Max Born in 1926. It is the square of these amplitudes that has direct physical meaning according to Born's interpretation. The probabilistic interpretation of quantum states is the cornerstone of the Copenhagen interpretation of quantum theory.

Our explanation of the interference is a special case of a rather general law in quantum mechanics: If there are two or more ways for a quantum system to go from an initial state to some final state, and if each way has associated with it a probability amplitude A, then the total probability amplitude for that final state is just the sum of the amplitudes for each of the possible ways to get there

$$A_{\text{final}} = A_{\text{path1}} + A_{\text{path2}} + A_{\text{path3}} + \dots.$$

The sum is a reflection of our ignorance: If we cannot know which path will be taken by the system, then the amplitudes associated with all possible paths must be summed. Only in such cases where there is an inherent ambiguity as to the path a particle will take can interference occur. Again, interference is a wave-like phenomenon. Nevertheless, the reader may have noticed that the actual detection of an individual electron is as a point on the detector screen and is thus indicative of particle-like nature. There is no contradiction here: When the electron strikes the screen, there is no longer any ambiguity about its location. The wave function Ψ is spread out over the screen just before the electron strikes it. When it does strike, the wave function "collapses" the point where the electron strikes. This is an example of the collapse, or reduction of the state vector about which we will say more later. Quantum mechanics does *not* say that an electron is a wave, but rather that its probability of being found anywhere in space is determined from a wave function. When we speak of the particle nature of the electron, we mean that the electron is definitely either in one place or another or on one path or another. If only *one* path is available, then only one amplitude is involved, meaning there can be no interference. When we use the word "path" here, we do not always mean "path through space", but "path" in a more general sense. Examples of this will be seen in later chapters.

So what does all of this mean? The "explanation" of the interference effect has been given in terms of the addition of probability amplitudes, and this may seem rather abstract. Furthermore, the probability amplitudes seem to be mathematical entities rather than physical ones. In the case of waves on water or of light, we have a good physical picture for the origin of interference, as the waves in these cases consist of physical entities. But in the case of the electrons it is not even clear why it is that the probability *amplitudes* must be added rather than the probabilities themselves. By only adding probabilities, which are always positive numbers, it would just not be possible to obtain the cancellations required for destructive interference. In essence, this construction, where probability amplitudes are added, is forced upon us by the phenomenon itself—that is, by Nature's indifference to our classical way of thinking. The interference can also be interpreted in terms of the availability, or lack thereof, of information regarding the path taken by the electrons. Interference occurs only when both slits are open—that is, when there are two possible paths that the electron can take, but no way for the experimenter to know which path is actually taken. This leads to the rule that when one cannot know, even in principle, the path taken by the electron to arrive at the film, the probability amplitudes associated with the two possible paths must be added. This rule seems to be a fundamental law in the sense that it does not follow from some other law, as far as we know.

Of course, there are those who, justifiably, seek a more fundamental explanation. Of considerable interest is a whole class of theories, alternative to the standard quantum mechanics, known as *hidden variable theories*. Many, if not most, of these alternatives have features just as strange as standard quantum theory. Some types may not even be subject to experimental tests. However, some definitely are amendable to experiment and, in fact, have already been subjected to experimental tests. We shall discuss such tests in Chapter 5.

Getting back to the double-slit experiment with electrons, the alert reader may have questioned what we just said about there being no way to know the electron's path when both slits are open. In the case of the bullets it is not hard to devise a way to track every bullet as it leaves the gun, proceeds through either of the slits (or neither, for that matter), and embeds itself in the wooden screen. This could be done with a camcorder. The recording could be played back in slow motion so that each bullet could be tracked. There is an implicit assumption being made here that the light that continuously reflects off the bullets and into the camera has no effect whatsoever on their motion. This is valid because the energy and momentum of the visible light-waves being reflected are very, very low compared with what would be required to change the motion of the bullet. But for electrons it is a different matter. Remember, electrons are point particles. How can we "see" point particles? Even if they were not point particles this would still be a problem. We see objects by the light they reflect, as in the case of the bullets. So, one might try to scatter radiation of a very short wavelength off the electrons. The shortest wavelength radiation available is gamma-rays. However, gamma-rays are very high in energy and momentum, and because the electron is a low-mass particle these gamma-rays can drastically affect the electron's motion. By placing a source of gamma-rays right behind the slits and observing the scattered gamma-rays, it might seem possible to detect the paths of the electrons. But their subsequent recoil changes everything about the experiment, since any particular electron will now hit the screen, if at all, in a place different than it would have without the gamma-ray source. The interference pattern will disappear even when both slits are open. Thus we have the following rule: If no "which-path" information is available we have interference, whereas if "which-path" information is available there is no interference. As we shall see later, it is not really even necessary for which-path information to actually be known by any observer in order to wipe out interference effects; rather, it is enough just for the information to be *potentially* known even if the observer does not do anything to obtain that information.

The inability to detect the electron's path, as described in the previous paragraph, is an instance of a rather general principle—namely, Heisenberg's *uncertainty principle*. The position of a particle and its momentum—essentially a quantity describing speed and direction—are, in quantum theory, called *incompatible quantities*. This means that we cannot measure each of them *simultaneously* with high precision. But the Heisenberg uncertainty principle means more than that. It means that incompatible quantities are *objectively indeterminate*. (We would have preferred that the Heisenberg uncertainty principle been called the *indeterminacy principle*—a more accurate description of its role in quantum theory.) Furthermore, if one of the quantities is known with high precision, then the other is known with very low precision. If the position of the electron is well known—for example, by the scattering of gamma-rays telling us which slit it has passed through—its momentum along a line between the slits is highly undetermined. This will wipe out the interference pattern. There are many pairs of incompatible quantities in quantum mechanics, and we shall encounter others later in the book.

The appearance of either the wave or the particle nature of the electron is strongly dependent on the experimental design. An experiment with only one slit open will detect only the particle nature of the electron, while an experiment with both slits open will reveal only its wave nature, these properties being mutually exclusive. In fact, we can say that the two types of experiments are mutually exclusive. It is important to note, however, that even in the case where the wave-like nature of electrons is exposed, the electrons are detected *individually* as localized particles—as spots on the screen or clicks of a detector. It is in the aggregate of an ensemble of electrons that we see the interference fringes. This is an example of what Niels Bohr called *complementarity*, meaning, in this case, that it is impossible to observe the wave and particle nature of electrons (and of other subatomic particles) simultaneously. In a particular experiment you can exhibit one of these attributes but not the other. The pairs of incompatible quantities related to the Heisenberg uncertainty principle are also complementary to each other in the way the term was used by Bohr. We shall encounter many other examples of complementarity in what lies ahead.

After reading the above, it might appear that some understanding has been obtained regarding the wave–particle duality of electrons. If so, this is an illusion. We really have not explained anything. What we *have* done is present a consistent framework for describing what has been seen in the double-slit experiment with electrons. Similar accounts can be given for other experiments carried out with other kinds of subatomic particle. Richard Feynman (a recipient of the Nobel prize for his work on quantum electrodynamics) once famously said that "nobody understands quantum mechanics". By that he meant that although it is possible to have a consistent mathematical framework for quantum mechanics that produces extremely accurate predictions, a deeper understanding still eludes us. Surely, one senses that there really ought to be a deeper understanding of what is really going on in experiments of the type we have described. And it is certainly the case that researchers have attempted to construct theories that provide a more satisfying picture of what really happens to an electron as it passes through the electron biprism of the Tonomura experiment. The problem is that some of these theories, referred to as *local hidden variable theories* (for reasons that are explained in Chapter 5), are really alternatives to quantum mechanics, and they fail to predict correctly the results of other kinds of experiments that can, and have been, performed in the laboratory. These experiments are in agreement with quantum mechanics and not the alternative theories (more about this will be discussed in Chapter 5). On the other hand, there is another kind of theory, called a *non-local hidden variable theory*, which turns out not really to be a truly different theory at all, but rather a reinterpretation of standard quantum mechanics. Such interpretations are not susceptible to laboratory tests, and furthermore have unsatisfactory features in their own right—features that are just as unsettling as the problems they are meant to solve.

So if you, the reader, are confounded by the double-slit experiment, you are in good company. But the double-slit experiment is perhaps the simplest example of nature not obeying the "common-sense" rules of the macroworld. Things become even more bizarre when two or more particles are in quantum states in which correlations exists between attributes of the particles. These states are called *entangled states*, and will be taken up later.

Can particles other than electrons show interference effects? The answer is "yes"—at least if the particles are not too big. Various kinds of interference experiments have been carried out with a variety of particles—most notably with neutrons from a nuclear reactor. Neutrons are about 2,000 times more massive than electrons, carry no electric charge as their name implies, and, though still very small, are not point particles. In fact, they seem to be made up of more elementary particles (the quarks) which themselves may be point-like. The diameter of a neutron is about 0.000 000 000 000 001 (10^{-15}) of a meter—unimaginably small. But in 1991, whole atoms, which are much bigger objects, produced interference patterns in an experiment performed by O. Carnal and J. Mlynek at the University of Konstanz. Atoms typically have diameters of about 0.000 000 000 1

(10^{-10}) of a meter—still not huge, but not point-like either. Even bigger objects, such as clusters of atoms, have been made to interfere. Quite recently, a group in Vienna headed by Anton Zeilinger has experimented with clusters of sixty carbon atoms, C_{60}, known variously as *buckyballs* or *fullerenes*, because their structure is reminiscent of the geodesic domes advocated as architectural designs by Buckminster Fuller. From the point of view of the microscopic scale, the C_{60} molecules are enormous, yet they can produce interference fringes. More recently, Markus Arndt and his group, also at the University of Vienna, have performed experiments in which very large organic molecules having up 430 atoms have displayed wave-like nature via quantum interference. They have shown that even more complex systems, having more than 1,000 internal degrees of freedom, can, if sufficiently isolated from environmental interactions, show nearly perfect quantum coherence. By "internal degrees of freedom" we mean the various types of motion, displacements, and rotations, and so on, that can occur within the molecule itself among its constituent atoms. As for the business about the "environment"—well, as the size of systems increase, the harder it becomes for experimentalist to isolate them from the "rest of the universe", the environment, so to speak, though we really mean other atoms or molecules that just happen to be near the system under study, the apparatus of the experiment, and so on. Interactions with these other systems tend to have the effect of destroying the quantum coherence that gives rise to quantum coherence effects, such as interference in the first place. The "decoherence" effects of environmental interactions will have much to do with the questions: Where is the divide between the quantum and classical worlds? Is there a divide? Why do baseballs not act like waves? We defer discussion to later chapters.

Bibliography

Arndt M., Nairz O., Vo-Andreae J., Keller C., van der Zouw G., and Zeilinger A., "Wave-particle duality of C_{60} molecules", *Nature* 401 (1999), 680.

Carnal O., and Mlynek J., "Young's double-slit experiment with atoms: A simple atom interferometer", *Physical Review Letters* 66 (1991), 2689.

Farmelo G., *The Strangest Man: The Hidden Life of Paul Dirac, Mystic of the Atom*, Basic Books, 2009.

Feynman R. P., Leighton R. B., and Sands M., *The Feynman Lectures on Physics*, Vol. III, Addison-Wesley, 1965. See especially Chapter 1.

Gamov G., "The Principle of Uncertainty", *Scientific American*, January 1958, p. 51.

Gerlich S., Eibenberger S., Tomandl M., Nimmrichter S., Hornberger K., Fagan P. J., Tüxen J., Mayor M., and Arndt M., "Quantum interference of large organic molecules", *Nature Communications* 2 (2011), 263.

van Heel A. C. S., and Velzel C. H. F., *What is Light?* McGraw-Hill, 1968.

Isaacson W., *Benjamin Franklin, An American Life*, Simon and Schuster, 2003.

Nairz O., Arndt M., and Zeilinger A., "Quantum interference experiments with large molecules", *American Journal of Physics* 71 (2003), 319.

Tonomura A., "Electron holography: A new view of the microscopic", *Physics Today* 43, 4 (1990), 22.

Tonomura A., Endo J., Matsuda T., Kawasaki T., and Ezawa H., "Demonstration of single-electron buildup of an interference pattern", *American Journal of Physics* 57 (1989), 117–120.

Waldman G., *Introduction to Light: The Physics of Light, Vision, and Color*, Prentice-Hall, 1983.

3

The Duality of Particles and Waves: Photons

3.1 The Symmetry of Nature

I don't know what a photon is,
but I know one when I see one.

ROY GLAUBER (recipient of the Nobel Prize in Physics, 2005, for his contributions to the quantum theory of optical coherence)

In the previous chapter we exposed the dual nature of small particles. We focused mostly on electrons—particles that are point-like and, therefore, the very essence of what it means to be a particle. An individual electron is always detected in a small, localized region on a scintillation screen, or seen as a track in a cloud chamber. However, when an electron passes through a screen with two slits, its fate will be determined by the interference arising from the two possible paths through the slits to the screen. The catch is that attempting to obtain information on which path the electron takes ends up destroying the interference altogether. The moral to be drawn from this is that if we set up an experiment to demonstrate the *particle-like* nature of the electron, surely the particle-like nature will present itself. Contrarily, if we set up an experiment to demonstrate the *wave-like* nature of electrons, we obtain exactly that, and thus we see that electrons do indeed seem to have "split personalities". The personality exhibited depends on the nature of the question (that is, the type of experiment) asked of them. Since it is the experimenter who ultimately decides what kind of experiment to perform, some have concluded that the universe, or at least a small part of it, is "observer-created". But there is a limited degree to which an experimenter can affect the behavior of a quantum system. Typically, experiments measure some particular physical attribute of a system where the outcomes are generally statistical. Other attributes of the quantum system under investigation may even be ignored. That is, we generally do not experimentally probe all aspects of the system in question, let alone of the entire universe. Thus we should be careful not to draw extravagant conclusions from the fact that the design of experimental set-ups can force different behaviors in quantum systems. However, within this limited range of possibilities the statistics of possible experimental outcomes will agree with the predictions of quantum theory.

When applying the scientific method during a standard scientific inquiry, it is generally held that an experiment performed under identical circumstances should produce the same results within experimental errors in measurement. However, in quantum mechanics that rule does not apply precisely. Rather, even as a matter of principle, the outcomes of the runs in a series of experiments performed with identically prepared quantum systems will generally be different. This difference has nothing to do with errors in measurements, but has everything to do with the inherent indeterminacy of the quantum world. The results from a compilation of the ensemble of runs of an experiment must be in accord with the statistical predictions obtained from the quantum-mechanical state vector. In this sense we must take a step back from the notion that "identically prepared experiments must yield identical results," and say instead that for quantum systems ensembles of identically prepared runs of experiments must yield the same statistical results. We shall provide a simple demonstration of this idea shortly.

But first, we ask the reasonable question: Is nature symmetric with regard to wave–particle duality? Given that particles such as electrons can somehow exhibit wave-like properties, is it possible that a wave-like phenomenon, such as light, can exhibit the kind of behavior we expect of particles? The answer to this question is "yes". The entity responsible for such behavior is the *photon*—an irreducible bundle, or quantum, of electromagnetic energy. The photon is certainly nothing at all like the corpuscles of light envisioned by Newton, nor can it be conceived as a particle in the same sense as can an electron. Unlike the electron, the photon has no mass, which in turn means that the position of a photon is not well defined. However, a photon can still have a well-defined direction of propagation, and it is this aspect of it that allows photons, excitations of the electromagnetic field, to behave like particles in an appropriate experimental arrangement. In the following section we will discuss such an experiment which leads us to an operational definition of the concept of the photon.

3.2 The Photon and Single-Photon Interference

In Chapter 2 we discussed a classical picture of light consisting of propagating electric and magnetic fields sinusoidally oscillating at right angles to each other. However, we mentioned nothing about the origin of light.

Consider an ordinary incandescent light. In the bulb there is a filament. When an electrical current passes through the filament, its high electrical resistance will cause it to heat up. With a sufficiently high current the filament can become so hot that it begins to glow. The higher the current, the brighter the filament will glow. The glow is due solely to the heat generated by the resistance in the filament, and is not the result of electrical discharge. But heat is a form of energy associated with the motion of atoms or molecules. When we measure the temperature of say, a gas composed atoms, we are taking a measurement of the average kinetic energy of the atoms of the gas. Kinetic energy is the energy associated with motion, and is one-half its mass times its velocity squared ($mv^2/2$). The greater the temperature, the greater the average energy of the atoms that make up the gas. Furthermore, as the atoms move about there are frequent collisions between them. If the energy is low, the states that the electron can occupy inside of the atoms will not be affected. But when the temperature becomes high enough the collisions can affect the internal states of the atoms, causing the electrons to be knocked from their lowest allowed energy states (ground states) into excited states. Once in an excited state, an electron very quickly jumps back to a lower state, eventually cascading back down to the ground state, where it may again become excited by another collision, and so on. Every jump of the electron to a lower level results in the emission of electromagnetic radiation. Let us represent two of the discrete energy levels of the atoms by the horizontal bars in Figure 3.1. The upper bar represents the state of higher energy, and the lower bar the state of lower energy. As we explained earlier, the electron will not stay in an excited state for very long—typically only about 10^{-8} seconds for low-lying states. Thus, it rapidly jumps, or makes a *transition*, to the lower level, and produces a bit of electromagnetic radiation as a result. This process, called *spontaneous emission*, is ultimately the source of almost all the light that reaches our eyes. This includes light from both natural sources, such as the Sun, and unnatural sources, such as incandescent and fluorescent lights, whether we see it directly from the source or by reflection. (There is one unnatural source that does *not* produce light by spontaneous emission: the laser, which we shall discuss shortly.) Now, when an electron jumps from a level of energy E_2 to lower energy level E_1, it loses the amount of energy represented by E_2–E_1, which is carried away by the emitted pulse of electromagnetic radiation. Note that when the electron reaches the ground state the atom stops radiating. The frequency of the light is determined by the relation

$$E_2 - E_1 = hf,$$

where f is the frequency of light (the number of cycles/second) and h is a very small number known as Planck's constant.* So, the energy difference of the atomic levels determines the frequency of the

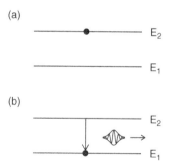

Figure 3.1 Two energy levels of an atom, where level 2 is the excited state and level 1 is the ground state. In (a) the electron is in the excited, whereas in (b) the electron has made a transition to the ground state and a photon has been emitted. If no photons are present at the moment the transition occurs, the process is called *spontaneous* emission.

---

* The numerical value of Planck's constant $h = 6.62618 \times 10^{-34}$ Joule/Hertz—a Joule being the unit of energy and Hertz the unit of frequency, 1 Hertz being one cycle (one complete wave crest-to-crest) per second.

emitted light. A high-energy difference results in high-frequency radiation such as X-rays and ultraviolet light. Correspondingly, a very-low-energy difference results in low-frequency radiation such as infrared light, radio waves, and microwaves. In between there is a very narrow band known as the visible-light spectrum. The human eye is not sensitive to light outside this band. Within the band, different colors correspond to light of different frequencies and wavelengths—red being the lowest frequency and highest wavelength, and violet being the highest frequency and lowest wavelength, detectable by the human eye.

Now, when we see the light emerging from an incandescent lamp we are seeing the effects of many atoms undergoing the process of spontaneous emission. Furthermore, generally speaking, many sets of transitions are involved to and from a variety of energy levels, resulting in the emission of many colors (frequencies) of light. As the temperature of the atoms is raised, the radiation emitted increases in overall intensity. However, the intensity is not uniformly distributed with respect to frequency. Instead, the distribution has a peak at some particular frequency—let us call it f_{peak}—and this frequency becomes higher as the temperature is raised (see Figure 3.2).

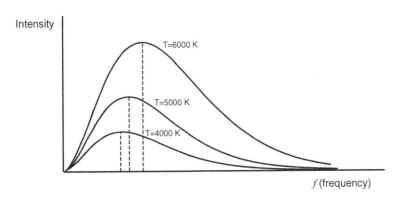

Figure 3.2 For an idealized object known as a *black body*, this is a plot of the intensity of light emitted as a function of the frequency for the body at three different temperatures. As the temperature is raised, the intensity increases overall and the peak intensity shifts to a higher frequency. This shift in the location of peak intensity with increasing temperature in Wien's displacement law.

The shifting of the intensity-versus-frequency curve as the temperature is changed is known as Wien's displacement law, which states that the frequency of peak intensity is proportional to the absolute temperature, mathematically written as $f_{peak} \propto T$, where T denotes the temperature in the Kelvin scale.[†] Wien's law tells us something about the temperature of a heated object: the frequency for which the emitted light is most intense will correspond to its color which, in turn, tells us the temperature. For example, a chunk of iron heated by a blow-torch will glow red. Wien's law tells us that since the iron is glowing red, it must have a temperature of about 2500 K because an object at that temperature will glow more intensely at that color than at any other. (It is also glowing quite intensely at frequencies outside the visible range—mainly in the infrared, which we can feel as "heat radiation".) Using Wien's displacement law we can determine the surface temperatures of distant stars. For example, the star Betelgeuse in the constellation of Orion is quite noticeably orange even to the naked eye, and from this we can deduce the surface temperature of the star to be about 3000 K. The Sun is a yellow star, indicating a surface temperature of about 6000 K. Every object in the universe having a temperature above absolute zero radiates in some part of the electromagnetic spectrum. People, and all living things for that matter, radiate light, though this light is not found within the visible-light spectrum: it is within the infrared range (longer wavelengths than those of visible light, with lower frequencies). In fact, military nightscopes function by picking up infrared light emitted by people and animals. Not only is everything *in* the universe radiating, but the entire universe *itself* is filled with radiation—the radiation left over from the Big Bang that created the universe some 14 billion years ago. The frequency of the peak intensity of this radiation is in the microwave part of the spectrum (lower in energy and frequency than those of infrared) and corresponds to an object of a temperature of only about 3 K. The universe has cooled to about 3 K as it has expanded to its present size. The energy contained in this *cosmic background radiation* is far greater than that given off by all the rest of the visible matter (stars, galaxies, and so on) of the universe combined.

[†] The Kelvin scale takes as its zero point absolute zero, $T = 0$ K (where K denotes Kelvin)—the temperature where all motion ceases, or would cease if the world were classical all the way down to the atomic level. Zero Kelvin is equivalent to $-273°$ C and $-460°$ F.

So, let us return to Earth and consider now only a single atom and a single pair of energy levels, as in Figure 3.1. When the electron is in the ground state, the atom does not radiate. But the lifetime of an excited atom is very short (around 10^{-8} seconds for optical transitions, as we have said), and eventually the electron jumps to the lower energy level, emitting a "pulse" of radiation of energy E_2-E_1 (the energy difference of the atomic levels) and frequency $f = (E_2-E_1)/h$. This burst of energy from the atom is the entity we call the photon—a "particle" of light. Also, a photon will emerge with particular polarization (vertical, horizontal, or at some angle as described for classical light-waves in the previous chapter), and we shall have more to say about photon polarization later.

We need to be extremely careful with regard to how we think of a photon as a particle of light. It is not a particle in the conventional sense, as is an electron. Electrons carry mass (though very small) and electric charge, and have whatever speed is given to them by any electric fields they encounter. Photons, on the other hand, are massless, carry no charge, and always travel at the speed of light. They should not be considered as anything like Newton's light corpuscles. Unlike those corpuscles, which would not be capable of diffraction or displaying any other wave-like characteristic such as interference, photons turn out to have a wave–particle duality, much like electrons. Just as in the case for electrons, the photon will behave like a particle if it is possible to determine *which-path* information. That is, in an experiment designed to elicit the particle-like nature of photons, the photons will oblige and behave just like particles. In fact, if we can perform some sort of which-path experiment on photons, we should be able to *force* them to act like particles if they really can have a particle-like nature. This, in fact, serves as an *operational* definition of what it means to be a particle: If it acts like a particle, then it is a particle—at least in a given experimental set-up. Conversely, an experiment designed to elicit the photons' wave-like nature will find them behaving just so. Thus, as with the electron, it will not be possible to carry out an experiment where both wave and particle natures are simultaneously on display. With regard to wave–particle duality, never the twain shall meet. This is, once again, a manifestation of Bohr's principle of complementary.

But what kind of experiments can be carried out with single photons? How can we be sure that we have only one photon at a time with which to experiment? In principle, one ought to be able to carry out an experiment of the type undertaken by Tonomura with electrons, where an interference pattern would be built up one photon as a time, as discussed in the previous chapter. The Tonomura experiment, in essence, was a Young's interference experiment per-formed with one particle (an electron) at a time. In fact, an experi-ment very much like this, but performed with light, was attempted as far back as 1909 by G. I. Taylor of Cambridge University. He was able to create interference fringes using "feeble" light—light weak enough from an energy standpoint that only one photon at a time, on average, should have been present between the source and the screen (a photographic plate). The "double slit" was actually a needle, and the interference fringes were seen in the diffraction pattern in the needle's shadow. The source of the feeble light was a gas flame, with a number of screens placed between the flame and the needle to reduce the intensity. In one run of the experiment a photographic film was exposed for about three months. Indeed, a diffraction pattern (inter-ference pattern) was visible on the film, apparently built up by one photon at a time.

Unfortunately, this experiment did not settle the matter of the wave–particle duality of photons. It was not until the 1950s that it was realized that thermal sources of light, such as a gas flame, are really not ideal as sources of single photons. Rather, it was discovered that thermal sources tend to produce photons in bunches of two. Most of the time, no photons emerge from the source; but when photons are emitted there is a high probability that there will be two of them. If one averages the energy emitted over a fixed time-interval, the *average* will be less than the energy contained in one photon. The average presents the illusion that only a single photon has been emitted in the chosen time-interval.[‡] But energy considerations alone are not enough to guarantee that there is, at most, one photon on average between the photon and the photographic plate, even when strongly attenuated. This kind of bunching effect, also known, after its discov-erers, as the Hanbury Brown–Twiss effect, cast doubt upon the

[‡] The photon bunching phenomenon and its effect on the average energy during a time interval is similar to the effect one billionaire would have on the average income on a street otherwise populated by paupers.

veracity of the Taylor experiment and other alledged single-photon interference experiments.

So, what other kinds of light source are there that might be suitable for generating single photons at a time? In 1960 a new source of light called the laser, or, more appropriately, LASER (Light Amplification by Stimulated Emission of Radiation) was invented. The photons in a laser beam tend to arrive randomly even when the beam is attenuated. This is an improvement over the bunching of photons seen in the thermal sources, but there still remains the possibility of having more than one photon present at any given instant.

Ideally, we should have a source of light that could emit photons in a fashion just the opposite of bunched—or *anti-bunched*. That is, the photons would come along in single file, evenly spaced in time. To obtain anti-bunched light we require a source consisting of just a few atoms—the best scenario being a single atom. The idea of using a single atom as a source of light would have been impossible to implement at the time of the Hanbury Brown–Twiss experiment in the 1950s. However, improvements in technology have made it possible to develop light-sources consisting of just one atom, though discussion of that technology is outside the scope of this book. Here we shall describe an experiment carried out in 1985 by Philippe Grangier, Gérard Roger, and Alain Aspect of the University of Paris, where, for the first time ever, single photons were used to create an interference effect. The experiment is considerably more subtle than the one performed with electrons described in the last chapter.

Let us first describe the nature of the light source, after which we will describe an experiment that demonstrates that single photons are indeed being generated. Grangier and collaborators used an atom of the element calcium. Figure 3.3 represents some of the energy levels available to one of the outer electrons of such an atom. The electron of the atom is elevated to the excited state by laser radiation. The decay of the atom proceeds in two steps. The electron first makes a transition to an intermediate state, emitting a photon of frequency f_1. Subsequently, the electron makes a transition to the ground state, and the atom emits a second photon of frequency f_2. The two photons fly off in different directions in order to conserve momentum (the atom itself will recoil when it emits the photons). In the experiment of Grangier and collaborators, both photons are used— one being the photon whose behavior is under investigation, and the

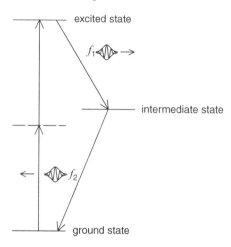

Figure 3.3 A cascade of atomic transitions to produce single photons. The atom is excited by the absorption of two laser photons, and then emits a succession of two photons on the downward transitions. The photons must fly off in different directions. The dashed line indicates a so-called *virtual atomic energy level*—a kind of intermediate energy level. Virtual levels are close in energy to real atomic energy level states, but unlike those, the virtual levels never become occupied by electrons. The electrons spend essentially no time in these levels.

other used as a trigger to alert some electronic circuits to get ready to detect the other photon. In their set-up, only photons emitted in opposite directions were used. We shall ignore these technical details in what follows as they are not relevant to what we wish to elucidate. Let us say that the photon of frequency f_1 is the one whose behavior we wish to scrutinize.

First, how do we know if we have only one photon? At this point we introduce an optical device known as a "beam-splitter". Typically, this is just a slab of glass, or some other semi-transparent material, placed at an angle of 45° to the light beam. The beam-splitter is a passive device (it does not create nor absorb photons), and merely transmits some of the light incident upon it and reflects the rest. The laser produces a beam of light containing many photons, wherein the average electric field oscillates sinusoidally, just as we would expect for a classical light-wave. So, when laser light falls onto a so-called 50:50 beam-splitter, the incident laser light intensity I splits it into two output beams, each of intensity $I/2$, as illustrated in Figure 3.4. It is important to understand that a beam-splitter only splits beams of light, and not the photons that compose the beam. That is, some of the photons will be transmitted and some will be reflected. In fact, for a 50:50 beam splitter 50% are transmitted and 50% reflected. Of course, for a classical light-beam, one does not need to think in terms of photons (discrete bundles of energy); rather, the beam is considered as a continuous entity being split into two continuous entities, like the water of a river flowing around an island.

Suppose now that only a single photon—the one of frequency f_1 obtained from the calcium atom—is incident on the beam splitter. Imagine we have the same kind of set-up as in Figure 3.4, but that we now have detectors sensitive enough to detect the presence of just one photon placed in each of the output beams, as shown in Figure 3.5. Also shown is a correlator which looks for coincident counts in the outputs of the detectors placed in the transmitted and reflected beams. Because the beam-splitter can only transmit or reflect the photon, only one of these detectors should "click", thus registering a single photon if there is truly a single photon incident upon the beam-splitter. There should be

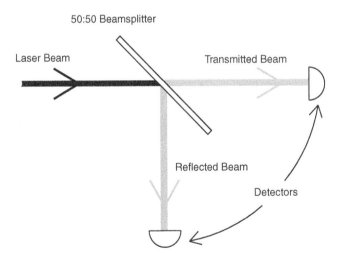

Figure 3.4 A 50:50 beam-splitter transmits half of an incident laser beam, and reflects the other half. The photons themselves are not split by the device. Half of them are transmitted and half are reflected.

no coincident clicks of the two detectors. Over a large number of runs of the experiment, each of the detectors should click alone in 50% of the trials, and again, there should never be simultaneous clicks of the detectors. The photon detections should therefore be *anti-correlated*. This is exactly what Grangier and his colleagues saw in the first part of their experiment. In fact, if there were simultaneous clicks in the detectors during the experiment, it would indicate that the beam-splitter was actually splitting photons somehow, or that the source was really producing more than one photon. There are, in fact, *active* optical devices that can split photons, and we shall describe such devices in the next chapter. But again, the beam-splitter splits beams, not photons, so in that sense it is a *passive* device.

Notice that the experimental set-up sketched in Figure 3.5 is a which-path kind of experiment, and therefore illustrates the particle nature of photons. Of course, that is just what Grangier and collaborators wanted to show. Nevertheless, there is still within this experi-

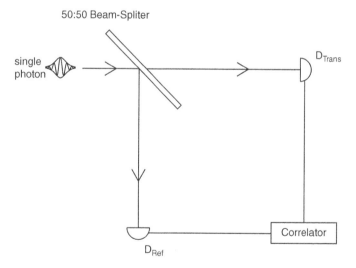

Figure 3.5 A single photon incident on a 50:50 beam-splitter. There is a 50% chance that a photon will be transmitted and a 50% chance it will be reflected, but it cannot be determined what will happen to any particular photon. The behavior of the incident photon is truly random. Placing detectors in the reflected and transmitted beams yields the particle-like nature of the photon, and which-path information is obtained.

ment an example of indeterminacy: we cannot predict which path will be taken by any given photon. Quantum theory predicts the probability that a photon will be transmitted, or reflected. If the beam-splitter is 50:50, then in half of an ensemble of runs of the experiments the photons are transmitted, and in half of the ensemble the photons are reflected. That is what is observed experimentally.

In this experiment the photons act as if they are particles, in the sense that they are in one output beam of the beam-splitter or the other. There is no ambiguity of path, as the firing of the detectors— one or the other but never both—provides which-path information. The output beams would have to come together somehow in order to create interference. To accomplish this, we need first to remove the two detectors and then direct the output beams of the beam-splitter onto opposite sides of yet another beam-splitter. This scrambles the path information and creates the ambiguity required for interference. The detectors are now sitting at the outputs of the second beam-splitter, as pictured in Figure 3.6. This setup is called a *Mach–Zehnder interferometer* (MZI). The box labeled by the angle θ in the upper arm is a device that allows for slight adjustments of the path length in the clockwise path—an adjustment known as "phase shift". Beam-splitter BS2, like BS1, is a 50:50 beam-splitter. The click of either detector reveals no which-path information. A photon taking the clockwise path can be transmitted to D1 or reflected to D2. Likewise, a photon taking the counter-clockwise path could be transmitted to D2 or reflected to D1. In writing the last two sentences we have assumed that each photon takes a definite path—either clockwise or counter-clockwise between the beam-splitters. If such were truly the case we would expect, in a long series of experimental runs, with a single photon injected as indicated in Figure 3.6, that each of the detectors D1 and D2 would click in 50% of those runs; but generally this does not happen. By changing the relative path-lengths of the interferom-eter (which amounts to changing θ) it is possible to change the rates of the clicking of the detectors such that the rates oscillate as path-length difference is changed, as shown in Figure 3.7. Note that at certain path-length differences the count rate in one detector goes to zero, but goes to a maximum in the other. The oscillations in the count rates are nothing but interference fringes, and the fringes of D1 are out of "phase" with those of D2. (Note that at *some* path-length differences the count rates of the two detectors *are* the same.)

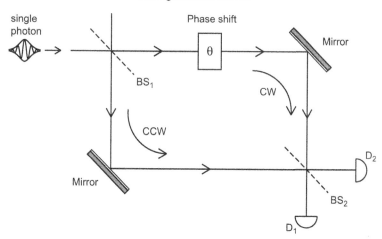

Figure 3.6 A Mach–Zehnder interferometer with a single photon entering the first beam-splitter from one direction and no photons from the
other. Interference is present in the output beams of the second beam-
splitter, due to the lack of information on which path is taken by the
photon. This device displays the wave-like nature of photons. The phase
shift represents the difference in the lengths between the clockwise (CW)
and counter-clockwise (CCW) paths.

Regardless of the path-length difference, the sum of the outputs of
the two detectors is constant: all the photons must go somewhere!
But how do the interference fringes arise?

Recall that quantum interference can occur whenever which-path
information is not available. The function of the second beam-splitter
is to ensure that such information is not available: it scrambles the
information in such a manner that when a detector clicks there is
simply no way of knowing whether the photon followed a clockwise
or counter-clockwise path. This means that the path taken by the
photon is objectively indeterminate. People sometimes say that the
photon seems somehow to have taken *both paths simultaneously*, although
such a statement should not be taken literally. The point is that the
unavailability of which-path information makes possible the wave-like
nature of the photon so that the interference can be seen at the level
of a single photon.

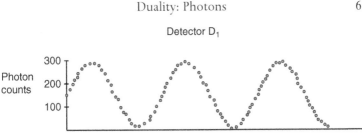

Detector D$_1$

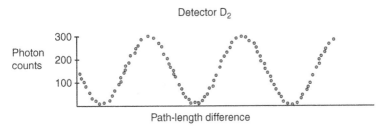

Detector D$_2$

Figure 3.7 The photon counts of the two detectors at the outputs of the Mach–Zehnder interferometer as functions of path-length difference. The oscillations are single photon interference fringes. Note that the fringes in the two detectors are out of phase with each other.

We can understand the origin of the interference in this experiment symbolically as follows: Let $|\text{CW}\rangle$ represent the state with the photon traveling in the clockwise path, and let $|\text{CCW}\rangle$ represent the state with the photon traveling in the counter-clockwise path. As seen in Figure 3.6, a photon enters the first beam-splitter from the left, while only the vacuum (that is, no photon) enters from the top. On the other side of the beam-splitter the quantum state is given by

$$|\text{After BS}_1\rangle = \frac{1}{\sqrt{2}}(|\text{CW}\rangle + i|\text{CCW}\rangle),$$

which is a superposition of the states where the photon propagates clockwise and counter-clockwise around the interferometer, with a 50% chance that the photon is going clockwise, with probability $P_{\text{CW}} = |\frac{1}{\sqrt{2}}|^2 = \frac{1}{2}$, and a 50% chance that it is going counter-clockwise, with probability $P_{\text{CCW}} = |\frac{i}{\sqrt{2}}|^2 = \frac{(-i)(i)}{2} = \frac{1}{2}$. The factor i appears in front of the counter-clockwise photon states $|\text{CCW}\rangle$, because that photon is reflected into that CCW beam by the beam-splitter, and because reflected light always picks up such a factor, which corresponds to a phase shift of 90°. In some sense there is nothing quantum-mechanical about this phase shift: it happens for purely classical beams undergoing a reflection. Placing the i in front of $|CCW\rangle$ is how we take this shift into account at the level of a single photon, which has to be described via quantum mechanics. Because the meaning of superposition states cannot be overemphasized, it is important to understand that the photon described by the state $|After\ BS_1\rangle$ has no objectively definite path around the interferometer. It is not correct to say that it is in either the CW beam or the CCW beam. It really is not the case that it is in both beams either: its beam location is indeterminate. This is just as in the case of the double-slit experiment with electrons.

The CW beam is subjected to a phase shift (a change of path length relative to the CCW beam) as represented by the angle θ. If d represents the difference in the lengths of the two paths around the interferometer, the angle θ, in degrees, can be written $\theta = kd$, where the constant $k = 360°/\lambda$, λ being the wavelength of the light. If the path lengths are the same, $d = 0$ and thus $\theta = 0$. The complete single-photon state inside the interferometer after the phase shift, but before the second beam-splitter, is

$$|After\ phase\ shift\ \theta\rangle = \frac{1}{\sqrt{2}}[e^{i\theta}|CW\rangle + i|CCW\rangle].$$

The exponential quantity $e^{i\theta}$ is an example of what electrical engineers call a *phasor* (not to be confused with the weapons of that name from *Star Trek*!). It is equivalent to $\cos\theta + i\sin\theta$. The complex conjugate of $e^{i\theta}$ is $e^{-i\theta}$, which is equivalent to $\cos\theta - i\sin\theta$.

The quantity $e^{i\theta}$ can be represented by a circle of radius 1 if we plot $\sin\theta$ versus $\cos\theta$, as seen in Figure 3.8, where we also give the standard definitions of the sine a cosine functions in terms of a right triangle. The phase shift θ is represented by a point on the circle.

Finally, the second beam-splitter, BS_2, again taking into account photon reflection, causes the transformations

$$|CW\rangle \rightarrow \frac{1}{\sqrt{2}}(|D_1\rangle + i|D_2\rangle),$$

$$|CCW\rangle \rightarrow \frac{1}{\sqrt{2}}(|D_2\rangle + i|D_1\rangle),$$

a)

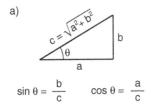

$$\sin \theta = \frac{b}{c} \qquad \cos \theta = \frac{a}{c}$$

b)

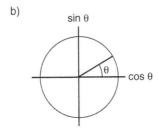

Figure 3.8 (a) Definitions of the sine and cosine functions of the angle θ of a right triangle. (b) A graph of $\sin \theta$ versus $\cos \theta$ representing the phasor $e^{i\theta}$. The circle is of radius 1, and the phase shift angle θ is represented by points on the circle, or the direction from the center to the points, as indicated.

where $|D_1\rangle$ and $|D_2\rangle$ represent the states where the photon is directed toward detectors D_1 and D_2, respectively. Thus, the full state after the second beam-splitter is

$$|\text{After BS}_2\rangle \rightarrow \frac{1}{2}[(e^{i\theta} - 1)|D_1\rangle + i(e^{i\theta} + 1)|D_2\rangle].$$

The probability that detector D_1 clicks—that is, detects the photon— is given by

$$P_{D_1} = \frac{1}{4}(e^{-i\theta} - 1)(e^{i\theta} - 1),$$
$$= \frac{1}{2}[1 - \cos \theta],$$

while similarly, the probability that D_2 clicks is

$$P_{D_2} = \frac{1}{2}[1 + \cos \theta].$$

Note that the two probabilities add up to 1—that is, $P_{D_1} + P_{D_2} = 1$—as they must. These probabilities oscillate as the difference in the path-lengths of the interferometer, which in turn changes θ. These oscillations are the interference fringes seen in Figure 2.7. The *origin* of the fringes stems from what happens at the second beam-splitter: BS$_2$ mixes the two beams and essentially scrambles the information on which beam the photon takes to get to it. A change of path-length difference modulates, through the phase shift θ, the propensity for the photon to be transmitted toward one detector or the other. If the path-lengths are identical then $\theta = 0°$, and because $\cos 0° = 1$, the probability that D$_1$ clicks is zero, that is, $P_{D_1} = 0$, while the probability of that D$_2$ clicks is 1, $P_{D_2} = 1$. So, with equal path lengths in the interferometer, detector D$_2$ clicks in 100% of the runs while D$_1$ never does. If the path-length difference was such that the phase shift angle was $\theta = 180°$, the behavior of the detectors would be reversed, because $\cos 180° = -1$. At intermediate phase-shift angles, both detectors will click but will do so in proportion to the probabilities for that particular phase shift. In these cases, unlike the special cases for $\theta = 0°$ or $\theta = 180°$ in which one knows that the photon will always go in the same direction after the second beam-splitter, it is not possible to predict the direction taken by any given photon after it leaves that beam-splitter. In the case where $\theta = 90°$, the probabilities for either detector clicking are, because $\cos 90° = 0$, identical and equal to $\frac{1}{2}$, meaning that the photons that emerge for the second beam-splitter are equally likely to go toward either detector. The fascinating thing about this is that at the most fundamental level, apart from those special cases just mentioned where one or the other detector clicks with certainty, the statistical predictions of quantum mechanics reflects the fact that chance is an irreducible element of nature—that there is no deeper explanation behind those statistical predictions. At least, that is the point of view of the Copenhagen interpretation. We shall have more to say about this later.

In summary, we have learned that photons will act like particles if an experiment is set up to reveal which-path information, whereas photons will have a wave-like character—that is, they can exhibit interference—if which-path information is prevented from being revealed. These complementary behaviors exactly parallel those of electrons and other particles. Especially note that interference involving photons is seen in the *aggregate*, not in the individual photons, just as in the case of electrons. In experiments such as Young's, a classical explanation of interference is possible, which is precisely what we provided in the previous chapter. In cases like that, the light-beam contains many photons simultaneously (and that would also be true of a laser beam, unless strongly attenuated by darkened glass), though we do not need to even invoke the photon concept. But the experiment of Grangier involved only one photon at a time to produce interference effects, and it is at the single-photon-at-a-time level where the interpretational difficulty arises in understanding the origin of the interference—just as in the case of electrons. Ultimately, all interference effects have this same explanation at the level of single particles arriving one at a time.

3.3 Delayed-Choice Experiments

In the single-photon experiments just described, a choice has to be made in advance as to which kind of experiment is to be performed. To obtain the particle nature of a photon, detectors must be placed at the outputs of just the one beam-splitter; to obtain the wave nature, the outputs must be fed into another beam-splitter. But suppose that somehow the choice of experiment performed could be delayed until *after* the photon goes through the first beam-splitter. Can we perhaps force it into acting wave-like or particle-like by changing the experiment in an instant just before the photon is detected somewhere? Obviously, given the enormous speed of light, it would not be possible to rapidly rearrange the apparatus in order to interchange the two arrangements described above. However, remember that the essence of this whole business is the availability, or lack thereof, of which-path information, so *any* way that path information can rapidly be revealed or hidden serves the same purpose as rearranging the apparatus. A technically feasible way of performing a delayed-choice experiment is sketched in Figure 3.9. The device, denoted PC in the counter-clockwise beam, is a *Pockels cell*—an electro-optical gadget that is connected to an electrical circuit containing a very fast switch. The photon injected into the interferometer at the first beam-splitter will have a definite polarization. The Pockels cell consists of a crystal across which a voltage may be applied. If the voltage is off (if the switch is open), the photon passes through undisturbed and the interferometer demonstrates the wave

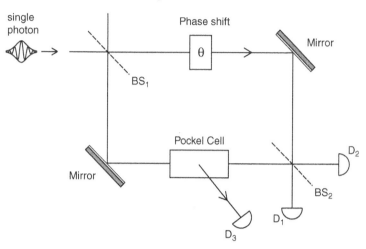

Figure 3.9 Schematic for a delayed-choice experiment. A Pockels cell connected to a fast switch is placed in one of the arms of the interferometer. When the cell is off, no path information is available, and single photon interference will occur. But when it is on, even after the photon has passed though the first beam-splitter, no interference is present in the output due to the availability of path information.

nature of photons, because no which-path information is available. But if the switch is closed, the voltage applied across the cell causes the polarization of the photon to rotate, and with the help of a polarizing filter is deflected towards detector D_3. In this case, if D_3 clicks, we *know* the photon has taken the counter-clockwise path. But even if it does not click, we still obtain the path information as long as the switch is on. To have reached either D_1 or D_2, the photon *must* have gone around the clockwise path. In such cases the detectors D_1 and D_2 would click randomly as a result of the random transmission or reflection from the *first* beam-splitter, and then randomly again from the *second* beam-splitter. Inasmuch as the second beam-splitter reflects or transmits the photons randomly, exhibiting no interference when the Pockels cell is on, that beam-splitter has effectively been removed.

The delayed-choice experiment was proposed in 1978 by John Wheeler at Princeton University. In the mid-1980s, two groups of experimenters—one headed by Carroll Alley of the University of Maryland, and another headed by Herbert Walther at the University of Munich—performed delayed-choice experiments of the type just described. The results? Just what you might expect if you know something of the quantum world: When the Pockels cell is on, no interference is seen; when the cell is off, interference occurs. The experimenters confirmed that after the photons pass through the first beam-splitter, it is still possible to evoke particle-like or wave-like behavior by rapidly changing the type of experiment. To obtain a sense of how rapid the switching must be, it is worthwhile to provide some numbers. The path lengths of the interferometer were about 4.3 meters. With the enormous speed of light (3×10^8 meters/sec), a photon can travel the above path length in about 14.5 nanoseconds; that is 14.5 billionths of a second—a very short time indeed. But fortunately, a Pockels cell can be actuated in an even shorter time— less than 10 nanoseconds. This rapid switching made the delayed-choice experiment feasible in a laboratory setting.

In spite of the results, the Alley and Walther experiments just described actually present a bit of a problem. Both experiments used severely attenuated pulses of laser light, and that means, harking back to what we said about the nature of light-sources earlier in the chapter, that it is statistically possible for there to be two photons in the pulses of laser light rather than always just one, as was assumed by the experimenters. Most of the pulses probably did have only one photon, but the fact that there could occasionally be two is enough to render those experiments not exactly "clean" demonstrations of the delayed-choice experiment. However, in 2007 a French group headed by Philippe Grangier and Alain Aspect repeated the experiment using a source that produced, to a very high degree of certainty, single photons. They used a single atom trapped in a diamond crystal as the source. They also made use of a quantum random number generator (more about that in Chapter 6) to operate the switch which opens and closes the interferometer. Lastly, the time interval between the entry of the photon into the interferometer and the activating of the switch was much shorter than the time it would have taken a photon, traveling at the speed of light, to have reached the beam switch. The results of this very clean experiment are in agreement with the predictions of quantum mechanics.

It just so happens, though, that there is a way to perform the delayed-choice experiment on a much larger scale in both distance and time. This too was proposed by John Wheeler. The idea is to take advantage of a phenomenon known as *gravitational lensing*. According to

Einstein's general theory of relativity,[§] a gravitational field causes light-beams to be deflected towards the source of the field. This prediction was long ago verified, in 1919, by noting the apparent change in the positions of stars seen near the Sun during a total solar eclipse. The gravitational lensing effect is similar, but on a much larger scale. This idea is illustrated in Figure 3.10. Light-rays from a very distant astronomical object such as a quasar[**] are deflected gravitationally to an observer on Earth by a massive galaxy somewhere between the quasar and the Earth. The observer sees two images of the quasar on either side of the intervening galaxy, as shown in Figure 3.10.

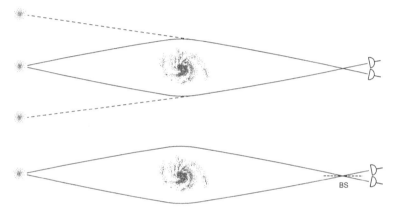

Figure 3.10 Wheeler's proposed delayed-choice experiment using a photon emitted by a distant quasar and focused toward the Earth by the gravitational lensing effect of a galaxy. The behavior of a photon that left the quasar perhaps a billion years ago can still be manipulated by the kind of experiment performed by the observer on Earth—either a which-path type of measurement (top), or an interference experiment (bottom).

[§] A theory of gravitation in which the gravitational force appears as a result of the curvature of spacetime, the curvature itself resulting from the presence of matter and energy.

[**] Quasar stands for quasi-stellar radio source—objects that are evidently galaxies in the early stages of their evolution. They are optically bright, but are also bright sources of radio waves. Normal stars are not bright radio sources. Quasars seem to be cosmological—that is, they lie at great distances and far back in time, millions or billions of light-years away (a light-year being the distance that light travels in one year—about 6 trillion miles). If we see a quasar 1 billion light-years away, we see it as it appeared 1 billion years ago, such are the great distances involved.

Gravitational lensing effects have, in fact, long been observed. (They constitute another confirmation of Einstein's general theory of relativity.) As is evident from the figure, the observations of the two images constitute a which-path measurement. But the experimenter is free to perform an interference experiment in which interference fringes will be observed (compare Figures 3.10 and 2.7, and note the similarities). But for now, suppose we consider only a single photon emitted from the distant quasar. Should the quasar be a billion light-years away, it will take a billion years for the photon to reach the Earth. That gives the observer a billion years to decide whether or not to perform a which-path experiment or an interference experiment. Of course, at the time of emission of the photon there were no human beings on Earth, nor were there any long after the photon passed by the intervening galaxy. The point is that the photon has almost reached the Earth before the experimenter makes his or her decision, and thus we have a delayed-choice situation with a delay of a billion years! Now, unless one believes that a decision made today can retroactively affect the behavior of a photon emitted a billion years ago in deep space, one would have to conclude that the properties of a photon are determined only at the moment a particular experimental arrangement is put into place, and that prior to this, the properties of a photon are indeterminate. This is the ultimate delayed-choice experiment.

3.4 Interaction Free Measurement

Finally in this chapter, we discuss yet another twist on interferometry with single photons. Conventional wisdom says that in order to detect the presence of an object, at least one photon must scatter from it. As should be of no surprise by now, conventional wisdom is sometimes wrong. By placing an object in one path of an interferometer, there are two ways to detect its presence—one being the case when it scatters a photon. The second way, as we show now, the object can be detected even when it does not scatter a photon. The approach we take originates from a proposal for *interaction free measurement* by A. C. Elitzer and L. Vaidman. The required set-up, pictured in Figure 3.11, is an interferometer with a single photon input at the first beam-splitter, where a solid object may or may not be present in the counter-

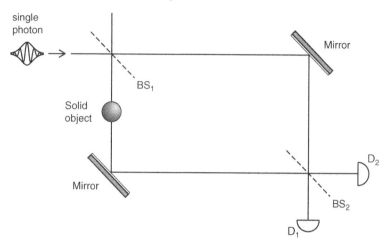

single
photon

BS$_1$

Mirror

Solid
object

Mirror

D$_2$

BS$_2$

D$_1$

Figure 3.11 Schematic for an interaction-free measurement. The presence of the object placed in the counter-clockwise path can be known even if no photons scatter from it through the altered counting rates of the two detectors, as explained in the text.

clockwise path. (In actual experiments that have been performed, the source of single photons was not atomic. In the next chapter we describe modern ways of producing single photons for such experiments and many others.) Let us assume for a moment that the object is not in the path, and that the relative lengths in the paths are adjusted so as to cause complete destructive interference in the output toward detector D1. This means that in every run of the experiment, without the object in the counter-clockwise path, only D2 clicks. Because of interference, the photon is always directed to detector D2.

But now let us assume that the object is placed in the counter-clockwise path, as indicated in Figure 3.11. This changes everything. Which-path information is now available. Referring to Figure 3.11, consider the following: At the first beam-splitter there is a 50:50 chance that the photon is transmitted horizontally (the clockwise path) or reflected downward (the counter-clockwise path). If reflected it will scatter from the object, and neither D1 nor D2 will click. We will know that the object is there, and that it scattered the photon. We know *that* because we *know* which path the photon has taken (the counter-clockwise), and we know *that* because it could not have taken the path (the clockwise) that would have caused either of the detectors to click.

If the photon is transmitted by the first beam-splitter, it will of necessity reach the second beam-splitter: it has nowhere else to go. Because one of the paths to the second beam-splitter is blocked by the object in question, no interference can occur, and the action of the second beam-splitter is just like that of the first: it will reflect or transmit the photon in 50% of the runs in the experiments where the photons take the clockwise paths. This means that detector D1 will occasionally click instead of D2. In summary, in half of the experimental runs, no detector clicks (that is, the object is there and scatters the photons), and in the other half D1 or D2 will click. In the runs that cause D2 to click (one fourth of the total number of runs), we learn nothing. Without the object in the counter-clockwise path, D2 will always click (remember that the interferometer is initially set up so that only D2 clicks), and it can also sometimes click if the object is in the counter-clockwise path. But for the remaining runs (the other one-fourth of the total), D1 clicks. The clicking of D1 unambiguously announces that the photon *has* taken the clockwise path, and that there is an object in the counter-clockwise path. (Remember that without the object in the path, interference prevents D1 from clicking.)

The effect is real, and has been demonstrated in the laboratory by Anton Zeilnger's group, at the University of Innsbruck at the time. Somehow the photons reveal the presence of an object that blocks a path even though they do not go near it. This "seeing in the dark" effect provides a hint towards one of the most unsettling of all oddities in quantum mechanics: *non-locality*, which refers to a property of some quantum phenomenon where actions have instantaneous effects at distant locations. Non-locality and its implications will be discussed in more detail in the following chapters.

Bibliography

Alley C. O., Jakubowicz A., Steggerda C. A., and Wickes W. C., "A delayed random choice quantum mechanical experiment with light quanta", in *Proccedings of the International Symposium of the Foundations of Quantum Mechanics, Tokyo*, ed. S. Kamefuchi (Physics Society of Japan, 1983) p. 158.

Elitzur A. C., and Vaidman L., "Quantum mechanical interaction-free measurement", *Foundations of Physics* 23 (1993), 987.

Grangier P., Roger G., and Aspect A., "Experimental evidence for a photon anticorrelation effect on a beam splitter: A new light on single photon interferences", *Europhysics Letters* 1 (1986), 173.

Hellmuth T., Walther H., Zajonc A., and Schleich W., "Delayed-choice experiment in quantum interference", *Physical Review A* 35 (1987), 25–32.

Jacques V.,Wu E., Grosshans F., Treuussart F., Grangier P., Aspect A., and Roch J.-F., "Experimental realization of Wheeler's delayed-choice gedanken experiment", *Science* 315 (2007), 966.

Kwiat P., Weinfurter H., Herzog T., Zeilinger A., and Kasevich M. A., "Interaction-free measurement", *Physical Review Letters* 74 (1995), 4763.

Taylor G. I., "Interference fringes with feeble light", *Proceedings of the Cambridge Philosophical Society* 15 (1909), 114.

Wheeler J. A., in *Mathematical Foundations of Quantum Mechanics*, ed. Marlow A. R., Academic Press, 1978, p. 9.

4

More Fun with Photons:
Photon-Splitting and its Uses

4.1 Active versus Passive Optical Devices

In the previous chapter we discussed the manipulation of a single photon of light by mirrors and beam-splitters for the purpose of demonstrating quantum interference using one photon at a time. We assumed throughout that the single photons in question were produced by spontaneous emission from an atom, where an excited electron randomly jumps to a lower atomic level and emits a photon in the process. This was the source of photons in the experiments of Grangier and collaborators. But it turns out that there are at least two reasons to seek other sources of photons. One is that during spontaneous emission the photons can uncontrollably fly away from the atom in random directions, with only the occasional photon being directed toward the experimental apparatus. Thus, there are issues of control and efficiency. The second is that quantum theory predicts phenomena even more exotic than those described in the last chapter if experiments can be performed with more than one photon.

The beam-splitter discussed in the previous chapter is a kind of optical device that we called "passive". As we emphasized before, the beam-splitter *splits* beams of photons, meaning that it will either reflect or transmit individual photons; but it will not alter them, change their frequencies, or create additional photons. In the ideal case, the beam-splitter does not destroy any of them either. On the other hand, there are optical devices that can not only destroy photons, but can create photons of a different energy and a different direction of propagation. Such optical devices are not constructed out of the ordinary *linear* materials, such as glass, used to make beam-splitters and mirrors. Rather, active optical devices are generally made from certain kinds of crystals, which we shall generically refer to as *non-linear* crystals. When a powerful enough laser beam is shone into the crystal, the strong electric field of the light causes electronic oscillations to occur. These oscillations are not only at the frequency of the incident light, but are also at frequencies that are higher or lower, and light is emitted in a highly directional fashion at these other frequencies. Light with strong electric fields is required to excite oscillations and create light at other frequencies. The study of these processes is known as *non-linear optics*. This field of research came into existence only after the invention of the laser in 1960, as only lasers produce light intense enough to cause electronic oscillations at frequencies different from that of the laser light itself. Because beam-splitters, lenses, and so on, do not emit light at any other frequencies, optics based on such devices is referred to as *linear optics*.

We consider here a particular non-linear optical process known as Type I *spontaneous parametric down-conversion* (SPDC). In Type I down-conversion, the emitted light all has the same polarization. In this process an intense laser beam of frequency f_L (f_L typically being in the ultraviolet part of the spectrum) enters a non-linear crystal. This intense beam is colloquially known as the "pump" beam. Inside the crystal, due to the presence of the strong oscillating laser electric field, two beams of frequencies f_1 and f_2 are generated and exit the crystal, as indicated in Figure 4.1. Notice that the laser pump beam exits the crystal along its original propagation direction, while the new beams propagate along angles to that beam. In fact, the two-dimensional picture in Figure 4.1 does not do justice to what actually happens in

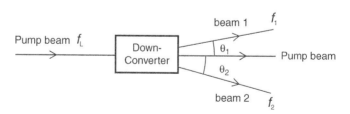

Figure 4.1 Light from the pump laser field of frequency f_L shone onto a down-conversion crystal is converted into two light beams of lower frequencies f_1 and f_2 where, from energy conservation, $f_L = f_1 + f_2$. The beams produced by down-conversion emerge at different angles as indicated. The beams produced by down-conversion are weak; most of the original laser beam light passes through the crystal undisturbed.

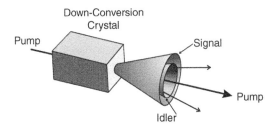

Figure 4.2 The light-beams produced by down-conversion actually emerge on opposite sides of the pump beam somewhere along two different concentric cones. The output beams are somewhat arbitrarily labeled signal and idler beams. We have shown only two cones, though in reality there is a continuum of cones.

such a process. In reality, photons created at frequencies f_1 and f_2 propagate out of the crystal along two concentric cones centered on the pump, as indicated in Figure 4.2. The origin of the cones is explained later in this section. The down-conversion process is not very efficient; the beams produced at the down-converted frequencies f_1 and f_2 are very low in intensity compared to that of the pump.

So far, we have described the down-conversion process only in terms of classical light-waves with intense fields. At the level of photons, here is what is happening. Occasionally, a photon in the laser pump beam (about one in a million) is split into two photons of frequencies f_1 and f_2, as shown in Figure 4.3, such that $f_L = f_1 + f_2$, as a consequence of energy conservation. (Recall that the energy of a photon of frequency f is given by $E = hf$, h being Planck's constant.) These "daughter" photons are created simultaneously. Because

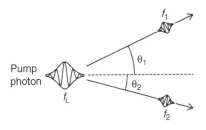

Figure 4.3 Down-conversion at the photon level. A pump photon is converted to two photons of lower frequencies. Only about one in a million pump photons undergoes a down-conversion.

conservation of momentum requires the total momentum of the two daughter photons to equal the momentum of the parent photon, the two daughter photons must emerge in the same plane that includes the direction of the pump photons, but the angles at which they emerge depend on the energies of the two photons. An example of this is given in Figure 4.1. If one imagines rotating beams 1 and 2 about the direction of the pump while keeping them opposite to each other, two concentric cones will be traced, as shown in Figure 4.2. Because a wide range of energies is allowed for the daughter photons, as long as $f_L = f_1 + f_2$, there will consequently be a continuum of concentric cones. This is illustrated in Figure 4.4 from the point of an observer looking into the spread of light emerging from the crystal. If the two daughter photons have different energies (say, one in the orange part of the spectrum and the other in the red) then the photons emerge along different cones opposite the center. Like symbols represent the exit points of the two paired, or conjugate, photons. On the other

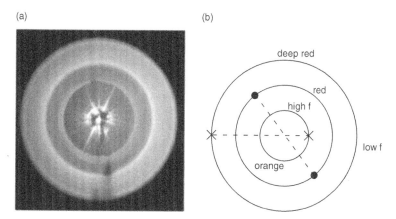

Figure 4.4 (a) A photograph taken from a head-on view of the cones of light emerging from the down-converter. The pump beam is at the center of the circles. The outer cones of light are the deep red, low frequency, while those near the center are in a somewhat higher frequency range corresponding to the color orange. (b) The dots and Xs connected by the dashed lines indicate the points where correlated photons will emerge. Note that the dots are actually on opposite sides of the same cone, meaning that the idler and signal photons will have the same frequency—half the frequency of the pump photons.

Source: Reproduced from National Institute of Standards and Technology.

hand, if the frequencies of the daughter photons are the same—that is, if $f_1 = f_2 = f_1/2$ then the paired photons emerge along the same cone but on opposite sides, as indicated by the symbols in Figure 4.5. Of course, for any given photon down-conversion, the exit points for a pair of daughter photons can be anywhere around the cones, though always opposite each other, as indicated by Figure 4.5. In a practical experiment we would want to select photons of the same frequency (and energy) emerging from two locations on opposite sides of the circle—that is, propagating along the two directions within a pair—while ignoring all other pairs emerging around the circle. We can easily accomplish this by simply placing a beam block in front of the crystal with two holes drilled on opposite sides of a circle centered on the pump beam while the pump is also blocked. This will guarantee that only those photons coming through the holes are paired (created simultaneously), and all other photons, whether paired or unpaired, will be excluded. By this method we can create pairs of photons, correlated in the sense that they are born simultaneously. We shall assume in what follows that we are arranging to select photons of the same energy, as just described. It turns out that the selected photons will have the same polarization, which is perpendicular to the polarizations of the pump beam. We shall refer to these selected photons as "twins", as they are alike in every way except for their direction of propagation.

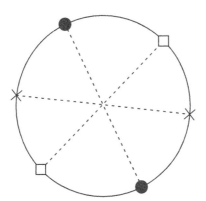

Figure 4.5 Correlated photons of the same frequency from opposite sides of a single cone can be taken from anywhere around the cone. Photons from the points connected by dashed lines are correlated.

How do we know that the daughter photons are created simultaneously? From experiment, of course. Back in 1970, David Burnham and Donald Weinberg (working for NASA) placed photodetectors in the output beams of a non-linear crystal pumped by a strong laser field. The distances of the detectors from the crystal were adjusted in such a way that when one detected a photon, so did the other (that is, simultaneous detector clicks). The *time* at which a photon *pair* is created via down-conversion is not known. The experiment indicates, however, that each member of the pair is created at the same time. This simultaneous creation of photons in a pair allows for some rather interesting applications of the down-conversion process, some of which will be discussed shortly. The 1970 Burnham–Weinberg experiment was key to much of the subsequent revolution in experimental quantum optics that took place throughout the 1980s and 1990s, and which continues to this day.

Before describing experiments performed with down-conversion devices, we reiterate that the rate of photon down-conversion is *very* low. Only about one in every million ($1: 10^6$) pump photons is down-converted into a pair of twin daughter photons. That is why most of the pump photons pass through the crystal. Other processes can occur as well. For example, two pump photons can be simultaneously down-converted into four photons producing twin two-photon states (two photons in each beam). The rate of conversion of this process is much, much lower than for producing just pairs out of one pump photon: roughly $1: 10^{12}$. One can have three pump photons down-converted into six, producing twin three-photon states, three in each beam, and so on, all occurring at lower and lower rates.

Before proceeding to new effects and new experiments, we should mention that all of the experiments on single photons discussed in the previous chapter can be, and have been, performed with the DC device replacing the single-atom electronic transitions. One can simply direct one of the down-converted photons to the first beam-splitter of an interferometer while its partner (twin) photon can be directed to a detector which, when activated, alerts the experimenter to the presence of the other photon (Figure 4.6). The single-photon interference experiment was redone by Paul Kwiat and Raymond Chiao in 1990 using down-converted photons. Later the "interaction free" experiment was performed by Paul Kwiat and collaborators using the same light-source. Note that only *one* of the paired photons

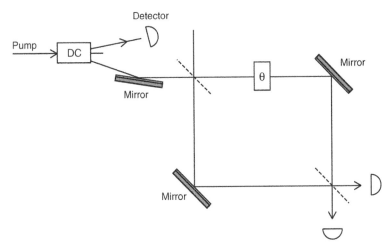

Figure 4.6 Down-conversion as a source of single photons to be used in an interference experiment. Because the photons are created in pairs, the detection of a photon in one of the output beams of the down-converter tells the experimenter that its twin photon has been injected into the interferometer.

produced is used directly in the experiment. But because the down-conversion produces tightly correlated (in time) pairs of photons, it is possible to perform new kinds of experiments where *both* photons are directly involved. These experiments, which we are about to describe, display much richer and more remarkable manifestations of quantum interference than can possibly be seen in single-photon experiments.

4.2 Two Photons on a Beam-Splitter

So far, we have considered experiments where only *one* photon falls onto *one* side of a beam-splitter, as shown in Figure 4.7. We know that the incident photon will be reflected or transmitted with probabilities of 50% for a 50:50 beam-splitter (henceforth we assume that all our beam-splitters are 50:50). But suppose another photon, identical in all ways to the first (same energy, same polarization) except for direction of propagation, is available and made to fall onto the beam-splitter from the other direction where there was previously no input light at all. This scenario is pictured in Figure 4.8. We then ask: What is the

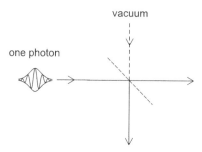

Figure 4.7 In the previous chapter we discussed experiments where a single photon fell onto a beam-splitter. The other input has only the vacuum.

nature of the output beams? There are two cases to consider having to do with the time of arrival of each of the photons at the beam-splitter. First, suppose that the photons reach the beam-splitter at different times. That is, the first photon is long gone before the arrival of the second; i.e. they do not overlap with each other. In this case, *each* photon will be transmitted or reflected 50% of the time, as usual. The photons "do not care" about each other, and act independently. In the second case, the photons strike the beam-splitter at the same time. This changes everything. There are now, in fact, four possible outcomes which are easy to enumerate. Referring to Figure 4.9, these are (a) both photons emerge in the "up" beam, $|0 \text{ up}, 2 \text{ down}\rangle$, (b) both emerge in the "down" beam, $|2 \text{ up}, 0 \text{ down}\rangle$, (c) both are transmitted, $|1 \text{ up}, 1 \text{ down}\rangle$, and (d) both are reflected, also $|1 \text{ up}, 1 \text{ down}\rangle$. For the

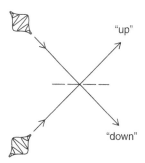

Figure 4.8 The case where single photons simultaneously fall onto both inputs to the beam-splitter. We label the output beams "up" and "down".

last two cases the output states are identical, and so one must add the probability amplitudes for these indistinguishable processes. But it turns out that the amplitudes are equal in magnitude but of opposite sign. This happens because in the case where each photon is transmitted through the beam-splitters the outcome state is just $|1$ up, 1 down$\rangle$, but when they are both reflected they each pick up a factor of $i = \sqrt{-1}$ such that the output state is $i \times i|1$ up, 1 down$\rangle = -|1$ up, 1 down$\rangle$ because $i \times i = -1$. Adding the amplitudes for the two indistinguishable ways to get the output state $|1$ up, 1 down$\rangle$ gives us $1+(-1)=0$. Thus the amplitudes from these two processes cancel each other out (a quantum interference effect), and the total output state is the superposition of the outputs of the two distinguishable processes:

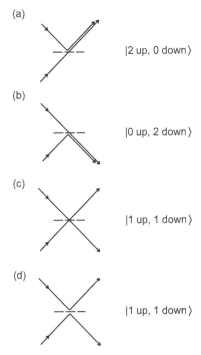

Figure 4.9 Possible outputs for two photons simultaneously falling on opposite sides of the beam-splitter: (a) both photons go in the up beam, (b) both go in the down beam, (c) both transmit through the beam-splitter, and (d) they are both reflected by the beam-splitter. The outputs of processes (c) and (d) are indistinguishable.

$$|\text{out}\rangle = \frac{1}{\sqrt{2}}[|2 \text{ up, } 0 \text{ down}\rangle + |0 \text{ up, } 2 \text{ down}\rangle].$$

The $1/\sqrt{2}$ factor appears because we have assumed a 50:50 beam-splitter, which means there is no preference as to which way they both go. If we put the photon detectors at the outputs of the beam-splitter, then when one "clicks" and the other does not, we can unambiguously distinguish between processes (a) and (b). There can be no quantum interference between distinguishable processes. Thus, if detectors are placed in the output beams, each should "click" in 50% of the runs of the experiment involving two photons hitting opposite sides of the beam-splitter simultaneously. There should be *no* coincident counts. This is reminiscent of the Grangier group's experiment involving only *one* photon. But here we need *two* photons coming from two different directions onto the beam-splitter, and the origin of the "no coincident counts" is two-photon quantum interference—an effect not operating in the single-photon case. We shall describe the relevant experiment momentarily, but first we want to say a few words about why the photons cluster together in the same beam upon leaving the beam-splitter.

In nature we recognize two fundamentally different kinds of particles: bosons and fermions. These kinds of particle have completely different personalities. Fermions, such as electrons, positrons, neutrons, protons and quarks, are "shy". They tend to be loners. In fact, they obey a very important physical principle known as the *Pauli exclusion principle*, which states that no two fermions can occupy the same quantum state at the same time. We should all be very grateful to Nature for the Pauli exclusion principle, for without it we would not be here. It is only because electrons obey this principle that chemical reactions can take place. Without it, all electrons of all atoms would be in the lowest energy states and none would be in the valence states, which participate in chemical reactions. The electrons fill up all the allowed energy states of an atom, and each state can contain only one electron due to the Pauli exclusion principle. Because the lower-energy levels are filled, some of the higher-energy electrons will be in valence states, meaning that they can be shared with other atoms and therefore form chemical bonds.

Bosons, on the other hand, tend to be gregarious. They like to be together, and they do not obey the Pauli exclusion principle. We can have as many bosons as we want in a single quantum state. The photon is a boson. Other bosons include the W and Z bosons that mediate the weak nuclear force, the gluons that mediate the strong nuclear force between the quarks, and the graviton which mediates the gravitational force. The fundamental difference between bosons and fermions partly accounts for what happens when two photons fall on opposite sides of the beam-splitter—they cling together. The clumping of photons is a version of the Hanbury Brown–Twiss effect that was discussed in the previous chapter. However, if two fermions were to fall on opposite sides of an appropriately constructed beam-splitter, they would emerge in different beams. This prediction of an anti-correlation effect is the fermionic version of the Hanbury Brown–Twiss correlation effect. Two experiments of this type involving electrons have been performed—one by Henny *et al.* at the University of Basel, Switzerland in 1999, and another by Kiesel *et al.* at the University of Tübingeu, Germany in 2002. Though they confirmed the effect, the electrons are particles that repel each other because they have the same electrical charge. Photons do not interact with each other directly as they carry no charge. To carry out an experiment analogous to one involving photons, one should use uncharged fermions. Neutrons are uncharged fermions. In 2006 an Italian group (Iannuzzi and collaborators at the University of Rome and other universities in Italy) performed an experiment with neutrons which cleanly demonstrated the anti-correlation effect.

Unfortunately, some of what we have just discussed about the gregariousness of photons cannot be generally used as a guide for how photons behave at a beam-splitter. If we consider the case where two photons each enter a beam-splitter from opposite sides (four photons altogether), as indicated in Figure 4.10, we might naively conclude on the basis of the previous paragraph that the output state should be

$$|\text{out}\rangle = \frac{1}{\sqrt{2}}(|4 \text{ up, } 0 \text{ down}\rangle + |0 \text{ up, } 4 \text{ down}\rangle).$$

But, alas, beam-splitters just do not work that way and this equation is wrong. Things are rather more subtle if more than two photons are involved. The actual output state will be

$$|\text{out}\rangle = \sqrt{\frac{3}{8}}(|4 \text{ up, } 0 \text{ down}\rangle + |0 \text{ up, } 4 \text{ down}\rangle) + \frac{1}{2}|2 \text{ up, } 2 \text{ down}\rangle,$$

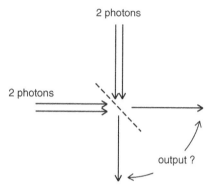

Figure 4.10 What happens when four photons, two at each input, simultaneously fall on the beam-splitter? As indicated in the text, the situation is more complicated than in the case with only one photon at each input.

meaning that in three-eighths of the runs, $\left(\left(\sqrt{\frac{3}{8}}\right)^2 = \frac{3}{8}\right)$, the state $|4 \text{ up}, 0 \text{ down}\rangle$ is detected, in another three-eighths the state $|0 \text{ up}, 4 \text{ down}\rangle$ is detected, while in one-fourth of the runs, $\left(\left(\frac{1}{2}\right)^2 = \frac{1}{4}\right)$, the state $|2 \text{ up}, 2 \text{ down}\rangle$ will be detected, and both detectors will click. States containing all four photons in the same output beam are still a major component of the total output state, but there is some probability that each beam could contain two photons. Notice, though, that it is still not possible for a single photon, or for three photons, to emerge in either output beam. The amplitudes of states containing one and three photons all cancel out. Although this four-photon state has been produced and detected in the laboratory, we shall not deal with it further here.

4.3 The Hong–Ou–Mandel Experiment

We now have some predictions. But what about experiments? In 1987, C. Hong, Z. Ou, and L. Mandel, working at the University of Rochester, used the two output beams of the down-conversion device in order to perform an experiment where single photons are allowed to fall onto opposite sides of a beam-splitter simultaneously. A sketch of the experiment is shown in Figure 4.11.

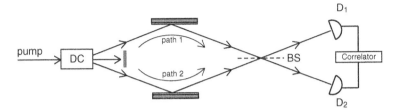

Figure 4.11 A sketch of the experiment of Hong, Ou, and Mandel. The pump beam is stopped and plays no role after producing the pair of photons in the down-converter (DC). The photons are reflected to a beam-splitter (BS) and then directed to detectors. The detector outputs are fed into an electronic correlator which searches for coincident photon arrivals (counts, or clicks) in the two detectors. According to the theory, if the photons arrive simultaneously there should be no coincident counts. The arrangement pictured is called the *Hong–Ou–Mandel interferometer*.

The idea behind it is as follows. Pump photons are down-converted into pairs, or "daughter" photons, as described previously. Remember, though, that the rate of the conversion of pump photons is very low. Most of the pump photons simply pass right through the DC crystal. But when a pump photon *is* actually down-converted, the daughter photons are produced simultaneously, and they also have the same polarization. That they have the same polarizations is of utmost is importance. The goal here is to get both photons (each emerging from the crystal in different beams, but otherwise indistinguishable) to opposite sides of the beam-splitter. This can be achieved with a pair of mirrors, as shown in Figure 4.11. However, if lengths of paths 1 and 2 are significantly different, each of the daughter photons will arrive at the beam-splitter at different times, and therefore each will act independently. That is, each will either be transmitted or reflected in 50% of the experimental runs. The outputs of detectors D1 and D2 are fed into an electronic correlator, which measures the number of coincident counts (when both D1 and D2 detect photons) within a small window of time. If one photon hits the beam-splitter even just shortly after the other so that the photons do not overlap each other, they could take different paths in which case both detectors would click. This would count as a coincident detection. However, if the path lengths are adjusted such that both photons hit the beam-splitter simultaneously, then, on the basis of the previous discussion, the number of coincident counts should go to zero. This is exactly what was found by Hong, Ou, and Mandel. By adjusting the position of the beam-splitter ever so slightly (on the order of just a few micrometers), the path lengths could be adjusted. The number of recorded coincident accounts in a 10-minute interval was plotted as a function of the position of the beam-splitter, as sketched in Figure 4.12. The sharp dip in the curve to nearly zero counts is the signal that the output state of the beam-splitter is

$$|\text{out}\rangle = \frac{1}{\sqrt{2}}(|2 \text{ up, } 0 \text{ down}\rangle + |0 \text{ up, } 2 \text{ down}\rangle).$$

In the regions to either side of the dip, the path lengths are sufficiently different in that each photon acts independent of the next, hence the high number of coincident counts. The important thing to remember is that the extinguishing of the coincident counting rate that occurs for two photons simultaneously falling on opposite sides of a beam-splitter is the result of the destructive interference between two indistinguishable processes (Fig. 4.9(c)(d)—a purely quantum mechanical result.

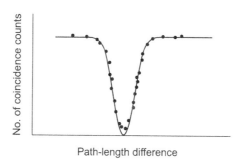

Figure 4.12 A sketch of the results of the Hong–Ou–Mandel experiment. By changing the lengths of one of the paths, the distance traveled to it by the photons also changes, and thus they can be made to arrive at different times. The number of coincident counts is plotted against the position of the beam-splitter. When the distances are approximately the same, the count rate takes a steep dive almost to zero. The rate does not go precisely to zero because of a difficulty in making the beams onto the beam-splitter precisely overlap. The extinction of the count rate to near zero can be used as a signal that both photons have traveled the same distance from the down-converter to the beam-splitter.

4.4 Variations on a Theme

The Hong–Ou–Mandel experiment is interesting in its own right, but beyond that it is an important starting point for a number of other experiments that further elucidate the nature of the quantum world. In this section we will consider two phenomena: the quantum eraser, and faster-than-light (superluminal) quantum tunneling.

4.4.1   The Quantum Eraser

Recall that earlier in this chapter we said that the paired photons obtained via down-conversion will have the same polarization. This means that each photon *within a pair* has the same polarization. Thus, the two photons are indistinguishable with respect to both energy and polarization. The fact that the photons are indistinguishable is what leads to the destructive interference between processes (c) and (d).

Now, suppose we do something to mark one of the photons to make it distinguishable from the other. The simplest thing we can do is rotate the polarization of one of the photons inside the interferometer, as indicated in Figure 4.13. In the extreme case we can rotate the polarization of one beam 90° to that of the other, giving maximum distinguishability. Nothing else about the photon is changed by

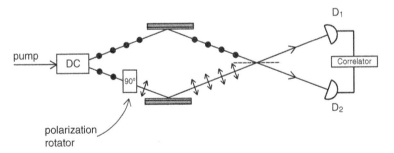

Figure 4.13 A Hong-Ou-Mandel interferometer with a polarization rotator in one beam. The dots along the beams indicated polarization perpendicular to the page. The rotator rotates the polarization of the lower beam by 90°, parallel to the page, as indicated by the arrows. The rotation of the polarization distinguishes the photons in the lower beam from those in the upper beam, and that destroys the interference that extinguishes the coincident count rate.

the rotator. Let us take the dots in Figure 4.13 to represent polarization perpendicular to the page, and the arrows to represent the polarization parallel to the page. We assume that the photons emerging from the down-converter are polarized perpendicular to the page, and further assume that the rotator rotates the perpendicular polarization into a direction parallel to the page polarization (a 90-degree rotation). Thus, when the photons strike the beam-splitter simultaneously, their distinguishability now causes each to act independently as if the other were not even there. So, in half the runs of the experiment, both photons will emerge in the same direction so that only one detector fires. In the other half of the runs they will go in opposite directions, and thus both detectors will fire. The latter case signals that the marking of one of the photons inside the interferometer has eliminated the quantum interference effect by removing the destructive interference that prohibits coincident photon counts. The beam-splitter itself is taken to be of a type that does not alter polarization (there are types that do), and the detectors do not measure polarization, but only tell of the arrival of photons. Now, we have seen before that the occurrence of quantum interference in multi-path processes is tied to information: lack of which-path information implies interference, while possession of which-path information destroys interference. But, in the present case, we go one step beyond these rules. The mere *potential* of possessing which-path information because we *can* determine which photon's polarization has been rotated, even if we do not actually do so, is enough to destroy quantum interference. Again, the polarization of the photons entering the detectors is never measured. It is very easy to determine that in about 50% of a large number of runs, only one detector or the other will fire, but not both. That is, each detector will fire alone in about 25% of the total runs of the experiment. In the *other half* of the total runs, both detectors will fire, and it is this double firing of the detectors that signals the lack of quantum interference. We have here another example of complementarity, but with a twist: *no* which-path information implies interference, whereas *potential* which-path information, even if never obtained by the experimenter, destroys it. If a 90-degree rotator is placed in *both* beams before the beam-splitter, the photons again are indistinguishable so that all the photon pairs again show interference.

An experiment of exactly the type just described was performed some years ago (1992) at the University of California at Berkeley by P. G. Kwiat, A. M. Steinberg, and R. Y. Chiao. They initially set up and "balanced" a Hong–Ou–Mandel interferometer. By balance we mean that that no coincident photon counts were observed, indicating that both paths from the down-converter to the beam-splitter were of equal length. Then, using a device called an optical half-wave plate, the polarization of one of the beams was gradually changed, whereupon the number of coincidence counts rose from zero to the maximum predicted when the polarization was rotated to up 90°—exactly in accordance with the predictions described in the previous paragraph.

Now comes the quantum-eraser part. Suppose we place in each beam emerging from the beam-splitter a polarizer oriented in the same direction, at 45° to the polarization of each of the original beams, as indicated in Figure 4.14. The photons emerging from each of these polarizers will be *identically* polarized, which has the effect of *erasing* the which-path information provided (or at least *potentially* provided) by the rotator. Note that the rotator is *inside* the interferometer, whereas the polarizers are *outside*. The predicted restoration of quantum interference, upon the erasure of which-path information, was observed by Chiao and his collaborators.

4.4.2 Can Light Travel Faster than Light? Quantum Tunneling

One of the more exotic predictions of quantum theory is "tunneling". If you take a tennis ball and throw it against a concrete wall, you

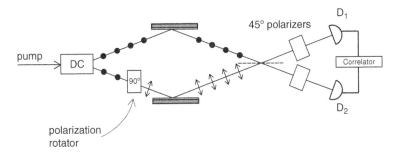

Figure 4.14 Placing 45-degree polarizers after the beam-splitter erases the distinguishing information provided by the 90-degree polarizer. Interference is revived, and the coincident count rate returns to zero.

would expect it to bounce back—a surefire prediction of *classical* physics. But if we were to perform a similar exercise with subatomic particles, say, by shooting electrons towards some kind of barrier, then, under certain conditions, it is possible for the electrons to "tunnel" through the barrier and be found on the other side in a "classically forbidden" region (Figure 4.15). If one shoots a number of electrons at the barrier, some will be reflected (perhaps most, depending on the nature of the barrier) and some will be transmitted. Those that are transmitted through have "tunneled". We should be careful here using the term "barrier". For electrons, the barrier results from the presence of an electrical field that acts to repel electrons. If the electron has enough kinetic energy (energy of motion) to overcome the repulsive force, the electron can simply pass over the barrier to the right, as shown in Figure 4.16(a). But if the kinetic energy is not sufficient to overcome the force, the electron can be either reflected to the left, or "tunneled" to the right with certain probabilities.

Tunneling is a strictly quantum mechanical effect; there is no classical analog, and, precisely because of this, it violates common sense and is yet another quantum mystery. We can describe it within the framework quantum mechanics (indeed, it was predicted within that framework!) but that does not remove the fundamental mystery

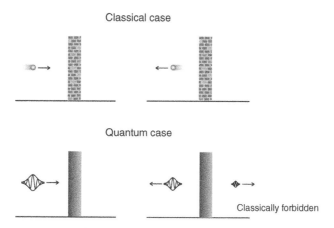

Figure 4.15 Tunneling. A ball—a classical object—thrown against a brick wall is always reflected; but a quantum-mechanical particle described by a wave function has a chance of tunneling through a barrier—a feat not possible for a classical particle.

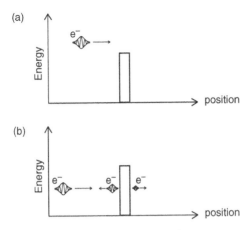

Figure 4.16 Electrons incident on an electric barrier represented by the rectangular box whose height represents the energy it contains. (a) If the electron has kinetic energy greater than the barrier energy, the electron can simply go over the top of the barrier. (b) If the electron's kinetic energy is less than that of the barrier, it can tunnel to the other side. There is a chance that it will be reflected and a chance that it can tunnel.

of tunneling any more than being able to describe the interference of material particles in a quantum framework (as we did in Chapter 2) removes the mystery of wave–particle duality. Tunneling plays an important role in many fundamental processes, and first appeared in quantum physics as a result of the work of Russian physicist George Gamov to explain a form of radioactivity known as alpha decay. Alpha decay is the process whereby a heavy nucleus such as uranium, which has 92 protons and 146 neutrons, occasionally spits out an alpha particle, which is just a helium nucleus consisting of 2 protons and 2 neutrons. The alpha particle tunnels through the potential energy barrier that exists because of the collective nuclear force due to the other protons and neutrons. The alpha particles that emerge do not have enough energy to have gone over the top of the energy barrier, indicating that they must be able to tunnel through it. Tunneling, like much that is quantum-mechanical, is probabilistic. Another example of a process where tunneling plays an essential role is that of thermonuclear fusion—the fusion process that powers the Sun.

Aside from the surprising nature of the tunneling effect itself, there is the question as to how much time it takes an electron (or any other particle) to tunnel through a barrier. The question cannot be addressed using classical physics, due to the lack of a classical counterpart. Quantum theory, however, seems to suggest that particles tunnel through barriers almost instantaneously. Historically, this has been difficult to test, at least with electrons. However, such a test *is* possible using photons: another application of the Hong–Ou–Mandel interferometer. It is possible because of the down-conversion process which produces two photons simultaneously, as previously discussed.

We will now describe an experiment on quantum tunneling performed at Berkeley by A. M. Steinberg, P. G. Kwiat, and R. Y. Chiao. Let us assume, as before, that the interferometer of Figure 4.11 is set up and balanced such that no coincidence accounts are observed. We now place in one of the beams a barrier—essentially a mirror, though not an ordinary mirror*—that reflects 99% of the incident photons from the down-converter, while 1% pass through, as shown in Figure 4.17. With the barrier in place, the interferometer is no longer balanced. In 99% of the experimental runs, only one detector fires, meaning only one photon is travelling beyond the beam-splitter. That photon has to travel in the lower beam in Figure 4.17 to reach the detectors. These runs are discarded. In the remaining 1% of the runs where both photons reach

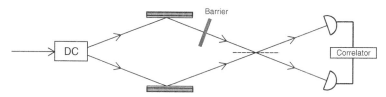

Figure 4.17 A Hong–Ou–Mandel interferometer with a barrier in one arm.

---

* The mirror consisted of alternating thin layers of two types of transparent glass through which light travels at different speeds. Each layer would only slow light down to different speeds, but with proper spacing, and with a multilayer metallic coating, 99% of the photons are reflect. In contrast, an ordinary mirror is glass with a metallic coating that can absorb up to about 15% of the light that falls on it.

the detectors (the photons arrive at different times), the experimenters were able to compare the arrival times between the tunneled photons and those that went via the other beam. In these runs, we have two possibilities: If the photons arrive at the beam-splitter at different times, and therefore act independently, there will be two sequential detector firings (sometimes the same detector, sometimes both), but with a delay between two firings. Because the beam-splitter scrambles the identities of the photons, it is still not possible to determine which arrives first. The other possibility is, of course, that the photons arrive at the beam-splitter simultaneously, and thus coincident counts are extinguished as a result of quantum interference, and only one detector fires.

With the barrier in place, the experimenters found that the photons no longer fell on the beam-splitter simultaneously. By changing the length of one of the paths, they could bring the interferometer back into balance where only one detector fires. Naively, one might expect that the photon *not* passing through the barrier should reach the beam-splitter first, and that the tunneled photon should have slowed down, as is usually the case for light passing through a medium of higher density. If that were the case, the interferometer should be brought back to balance by *lengthening the path without the barrier*, the lower path, just enough to delay that path's photon a little so that they fall on the beam-splitter simultaneously. But, alas, the experimenters found that *it was the path that contained the barrier that had to be lengthened* in order to bring the interferometer back into balance! It seemed that the photon in the path containing the barrier was traveling, not at the usual speed of light, $c = 300,000$ km/sec, but at about 1.7 times the speed of light—a *superluminal* speed! If the speed of light is the ultimate speed limit in the universe, as is generally supposed, how can light appear to travel faster than the speed of light?!

According to Einstein's theory of relativity, no signals can travel faster than the speed of light. This restriction is required by the notion of causality. If faster-than-light (superluminal) signaling were possible, there could be instances where events precede their causes and constitute violations of causality. Relativity is a classical theory, and in the everyday classical world we never see events preceding causes. But what about the quantum world? Is the experiment just described a harbinger of a new faster-than-light communications technology, perhaps like those "subspace channels" we hear about on *Star Trek*? Unfortunately, the answer is "no".

Signaling cannot be accomplished superluminally. There are, nonetheless, some *effects*, even in classical physics, that are superluminal, but they are not the result of signals moving faster than light. As a simple example, consider a laser that is being rotated such that its beam spot is moving across a screen, as pictured in Figure 4.18. If the laser rotates at a constant speed, the speed of the laser spot across the screen depends only on the distance of the screen from the laser. For the screen at a sufficiently great distance, the speed of the laser spot can exceed the speed of light—that is, it will be superluminal. But it is not possible for two people to communicate using this effect.

So, what is the explanation of the apparent superluminal tunneling effect revealed in the Chiao experiment? Recall from earlier that the two daughter photons generated via down-conversion from a pump photon are produced simultaneously. That said, we do not know when the pair *itself* is created; that is, there is uncertainty in the time of the pair creation. We can therefore think of each photon as being described by a "wave packet" spread out along the direction of propagation, as shown in Figure 4.19. These packets are really probabilities associated with the photon creation times. Because of strong correlations between the two photons, there are also strong correlations between the same parts of the wave packets. The probability of finding a photon is greatest at the peaks of the wave packet, but very small near the edges. Let the photon in the path with the barrier be photon 1, and let photon 2 be in the path without. Now, when the

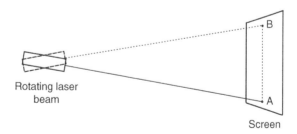

Figure 4.18 A rotating laser produces a moving spot on a distant screen from A to B. If the screen is sufficiently far from the laser, the speed of the spot across screen can exceed the speed of light. The motion of the laser spot cannot be used for communication.

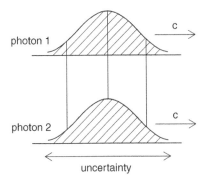

Figure 4.19 Wave packets (probabilities, actually) of two strongly correlated photons produced by down-conversion. The spread in the packets along the direction of propagation is associated with the uncertainty in the time that the photons are produced. The probabilities of finding the photons within the wave packet is greatest at the peaks.

wave packet of photon 1 reaches the barrier, the packet splits such that 99% is reflected and 1% is transmitted. The reflected wave packet looks like the incident, only moving to the left, while the transmitted one is much smaller and narrower overall, as pictured in Figure 4.20. Notice that the small wave packet of the transmitted wave packet is slightly ahead of the wave packet of photon 2. But the former is not traveling faster than the speed of light. Rather, the incident wave packet of photon 1 is thought to be "reshaped" as it travels, and the peak seen in *its* wave packet originated in the front end of the incident wave packet. Such reshaping effects have indeed been seen in other optical experiments. The upshot of the reshaping is that the peak of the transmitted wave packet is ahead of the peak of the wave packet of photon 2, and thus reaches the detector first. *Before the wave packet reaches the detector, the photon itself is neither in the reflected nor transmitted wave packets.* It is actually in a superposition of both, with a 99% chance of being in the former. On those occasions when the detector clicks, the entire quantum state of the photon (the aforementioned superposition) collapses onto the detector. Such collapses are thought to be instantaneous, or at least no experiment has so far shown them not to be. Such superluminal effects cannot be used for communications any more than can a rotating laser beam.

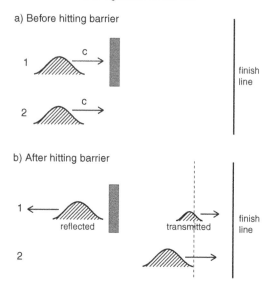

Figure 4.20 Both photons move at the speed of light. Photon 1 will encounter the barrier, while photon 2 has a direct path to the finish line. At the barrier, the wave packet of photon 1 is split, most of it being reflected, while the small part that is transmitted has a l peak just ahead of the peak of the wave packet of photon 2.

4.5 Controlling One Photon with Another: "A Mind-Boggling Experiment"[†]

We close this chapter with the description of an experiment performed by X. Zou, L. Wang, and L. Mandel at the University of Rochester in 1991. This experiment shows that it is possible to manipulate photon interference (to destroy or revive it) with actions on another photon that is not even in the interfering pathways. The experiment involves two down-conversion crystals in an arrangement, as sketched in Figure 4.21. Note that the downward-to-the-right output beams of both down-converters are aligned if the beam stop B (a moveable block) is not in the downward beam of the first down-converter (DC1).

[†] Quote in reference to this experiment taken from D. Greenberger, M. A. Horne, and A. Zeilinger, "Multiparticle interferometry and the superposition principle", in *Physics Today* 46 (1993), 22.

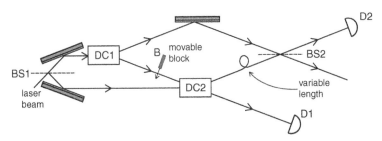

Figure 4.21 A sketch of the experiment of Zou, Wang, and Mandel. See the text for a complete explanation. Note that the downward output beams of the two down-converters are aligned and directed toward detector D1.

The experiment is perhaps best described by following through one run. We assume that the beam block B is removed from the beam, as indicated in Figure 4.21. An ultraviolet pump photon from the laser (an argon-ion laser) strikes beam-splitter BS1 (assumed 50:50), which produces a superposition state such that there is a 50% chance of being directed toward DC1 or toward DC2. Remember though, that because of the beam-splitter, the photon is not in either path, but in a superposition of both. When the pump photon encounters a down-converter, pairs of photons are produced. Note that a photon in the upward beam out of the beam-splitter will reach DC1 before the one in the downward beam reaches DC2, DC1 being closer to the beam-splitter. A pump photon reaching DC1 creates a photon pair, and one of those photons is directed to DC2. But a pump photon reaching DC2 can also create pairs of photons. Because of the fact that the pump photon is in a superposition state associated with the two output paths of the first beam-splitter, the outputs of the two down-converters will also be in superposition states. Note from Figure 4.21 that the downward output beams of the two down-converters are aligned and directed towards detector D1. Because of this alignment, when D1 clicks it is not possible to determine in which crystal the photon was produced. The uncertainty in the source of the detected photon along the aligned paths implies that the photon that reaches detector D2 is also of an uncertain source. That uncertainty is brought about by the action of beam-splitter BS2.

So, is this photon produced in DC1 or in DC2? Quantum mechanics does not allow us to provide a definite answer to this question. It is actually in a superposition of both possibilities, just as would be the case of a single photon exiting a beam-splitter, as described in Chapter 3. Thus, by changing the path length by, say, inserting in one of the beams pieces of glass of different thicknesses, or inserting optical fibers[‡] (as indicated by the loops in Figure 4.21) of different lengths, single-photon interference can be observed in detector D2.

If we now insert the beam stop B, it is evident that any photon detected by D1 *must* be produced in DC2. This, in turn, means that the photon reaching D2 was also produced in DC2. On the other hand, if D2 clicks but *not* D1, then we know that the photon had to be produced in DC1. In either case there is no ambiguity in the production of the photon, and there should be no interference at detector D2. This is exactly what Zou, Wang, and Mandel found. The significance of this experiment is that the only manipulations performed—the insertion or removal of the beam stop which destroys or revives interference—do not occur in the interfering pathways. In fact, this experiment illustrates that quantum interference is not merely interference from different paths, but from different processes.

In this chapter we have explored a number of experiments based on the production of pairs of photons in a non-linear crystal: the so-called down-conversion process. We are not yet finished with such devices, but before we continue we need to have a detailed discussion on the nature of *quantum entanglement*. In fact, we have actually already been using this concept in the last two chapters without explicitly saying so. We will now face this notion head-on, and confront the most unsettling aspect of quantum theory: *non-locality*.

Bibliography

Burnham D. C. and Weinberg D. L., "Observation of Simultaneity in Parametric Production of Optical Photon Pairs," Physical Review Letters *25* (1970), 84.

Chiao R., Kwiat P. G., and Steinberg A. M., "Faster than Light?", *Scientific American*, August 1993, p. 52.

[‡] An optical fiber is a thin cylindrical length of glass that carries light-signals. The light is kept inside the fiber by a phenomenon called *total internal reflection*. The reader has no doubt heard of optical fibers in connection with telecommunication and computer networks.

Henny M., Oberholzer S., Strunk, C., Heinzel T., Ensslin K., Holland M., and Schönenberger C., "The Fermionic Hanbury Brown and Twiss Experiment," Science 284 (1999), 296.

Hong C. K., Ou Z. Y., and Mandel L., "Measurement of subpicosecond time intervals between two photons by interference", *Physical Review Letters* 59 (1987), 2044.

Iannuzzi M., Orecchini A., Sacchetti F., Facchi P., Pascazio S., "Direct Experimental Evidence of Free-Fermion Antibunching," Physical Review Letters 96 (2006), 080402.

Kiesel H., Renz A., and Hasseltach F., "Observation of Hanbury Brown-Twiss centicorrelations for free electrons," Nature 418 (2002), 392.

Kwiat P. G., Steinberg A. M., and Chiao R. Y., "Observation of a 'quantum eraser': A revival of coherence in a two-photon interference experiment", *Physical Review A* 45 (1992), 7729.

Steinberg A. M., Kwiat P. G., and Chiao R. Y., "Measurement of the single-photon tunneling time", *Physical Review Letters* 71 (1993), 708.

Zou X. Y., Wang L. J., and Mandel L., "Induced coherence and indistinguishability in optical interference", *Physical Review Letters* 67 (1991), 318.

5

Entanglement and Non-Locality: Spooky Actions at a Distance

Superposition, in reality, is the only mystery.
R. P. FEYNMAN

5.1 The Only Mystery?

Well, it is and it isn't. The above quotation by Richard Feynman is quite correct if we are talking only about one particle at a time such as single electron that passes through a double-slit apparatus, as described in Chapter 2. The superposition is the sum of the probability amplitudes for a single electron to pass through either of the two slits. It turns out, however, that quantum mechanics has yet another mystery—one that occurs as a consequence of superposition states involving two or more particles or two or more beams of photons. We here refer to the concept of *entanglement*, also known as *non-separability*. The existence of entangled states is allowed, and even mandated, by quantum mechanics, and this leads us into yet another mystery: *non-locality*. In fact, we have already been using some of the consequences of entanglement in the previous two chapters without directly using the term. Here we will make the concept explicit. We warn the reader that, out of necessity, this chapter will be a little more technical than the previous ones, though no calculations are performed. In this chapter we number the equations as an aid to following the arguments.

To begin, let us temporarily regress to the use of the "quantum" coins introduced in Chapter 2. The state of a single coin could be heads, $|h\rangle$, tails, $|t\rangle$ (the only allowed "classical" states), or some superposition of the states, such as the balanced superposition

$$|\psi\rangle = \frac{1}{\sqrt{2}} (|h\rangle + |t\rangle). \tag{5.1}$$

We refer to the superposition as *balanced* because the probability of the coin being found heads (that is, measured to be heads) is $\frac{1}{2}$, as is the probability of it being found tails. Suppose now that we have two coins: coin 1 and coin 2. What are the possible states for the two coins taken together? Let $|h\rangle_1$ and $|t\rangle_1$ be the states for coin 1 and $|h\rangle_2$ and $|t\rangle_2$ be those for coin 2. For the *system* of the two coins, we can easily list all the combinations of "classical" possibilities as follows:

$|h\rangle_1|h\rangle_2$, meaning coin 1 is heads and coin 2 is heads;
$|h\rangle_1|t\rangle_2$, meaning coin 1 is heads and coin 2 is tails;
$|t\rangle_1|h\rangle_2$, meaning coin 1 is tails and coin 2 is heads;
$|t\rangle_1|t\rangle_2$, meaning coin 1 is tails and coin 2 is tails.

These states are called *product states* because we can specify the state of each coin independently and then multiply them together so that the state of the two-coin system can be written as

$$|\text{state of coin 1}\rangle|\text{state of coin 2}\rangle.$$

As quantum mechanics allows for superposition states, another possible state of the two-coin system is

$$\frac{1}{\sqrt{2}} (|h\rangle_1|h\rangle_2 + |t\rangle_1|h\rangle_2). \tag{5.2}$$

But it is easy to see that this state can be separated, or factored, into the form

$$\frac{1}{\sqrt{2}} [|h\rangle_1 + |t\rangle_1]|h\rangle_2, \tag{5.3}$$

where coin 1 is in a quantum superposition state, though coin 2 is definitely heads. The meaning of the two-coin state is that coin 1 exhibits objective indefiniteness with respect to heads and tails and has a 50% chance of being found either heads or tails, while coin 2 is always going to be found as heads. There are no correlations between the results obtained for the two coins. We have noted that the state of the two-coin system can be written as a product of the states of each of the subsystems; the states of the individual coins, as $|\text{state of coin 1}\rangle \ |\text{state of coin 2}\rangle$. Many more states (an infinite number, actually) of this sort are possible. For example, we could have

$$\frac{1}{\sqrt{2}}\left(|t\rangle_1|h\rangle_2 + |t\rangle_1|t\rangle_2\right) = \frac{1}{\sqrt{2}}|t\rangle_1[|h\rangle_2 + |t\rangle_2], \qquad (5.4)$$

where now coin 2 is in a superposition state, but coin 1 is definitely tails. For *any* two-particle system, where the state can be written in the form $|$state of particle 1$\rangle|$state of particle 2$\rangle$, and where the state of either particle could be a superposition of its available basic states, the system is said to be *separable*, meaning that each particle is entirely independent of the other. The state

$$\frac{1}{\sqrt{2}}\left[|h\rangle_1 + |t\rangle_1\right] \times \frac{1}{\sqrt{2}}\left[|h\rangle_2 + |t\rangle_2\right], \qquad (5.5)$$

where we have put in the multiplication sign for emphasis, is separable. Particle (coin) 1 is in a superposition, as is particle (coin) 2, but there are no correlations of any kind between the particles.

The notion of separability can be extended in an obvious way to systems of more than two particles. In fact, it can be trivially extended to states for any number of particles. However, separable states in general are not very interesting, as each particle is independent of all the others. On the other hand, multi-particle systems with states that are *not* separable are highly interesting indeed.

Consider the superposition state

$$\frac{1}{\sqrt{2}}\left(|h\rangle_1|h\rangle_2 + |t\rangle_1|t\rangle_2\right). \qquad (5.6)$$

Its meaning is as follows: The chance of finding *both* coins heads is 50%, and that of finding *both* coins tails is 50%. The coins are tightly correlated: either both will come up (will be measured to be) heads, or both will come up tails. We never get one heads and the other tails. Furthermore, it is not possible, quite unlike the previous cases, to write the totality of the two-coin state as a product of the states for each of the coins. Symbolically,

$$\frac{1}{\sqrt{2}}\left(|h\rangle_1|h\rangle_2 + |t\rangle_1|t\rangle_2\right) \neq |\text{state of coin 1}\rangle|\text{state of coin 2}\rangle, \qquad (5.7)$$

where, $\neq$ means "not equal". Our state is *not* separable; it is therefore called *entangled*. To see that the left side of eqn. (5.7) cannot be written as a product state, we can first assume that the state of each coin is a superposition of its heads and tails states:

$$|\text{state of coin 1}\rangle = a_1|h\rangle_1 + b_1|t\rangle_1,$$
$$|\text{state of coin 2}\rangle = a_2|h\rangle_2 + b_2|t\rangle_2, \tag{5.8}$$

where a_1, a_2, b_1, and b_2 are probability amplitudes. When we multiply them together, we have

$$|\text{state of coin1}\rangle|\text{state of coin 2}\rangle$$
$$= a_1a_2|h\rangle_1|h\rangle_2 + b_1b_2|t\rangle_1|t\rangle_2 + a_1b_2|h\rangle_1|t\rangle_2 + a_2b_1|t\rangle_1|h\rangle_2. \tag{5.9}$$

The "cross terms" $a_1b_2|h\rangle_1|t\rangle_2 + a_2b_1|t\rangle_1|h\rangle_2$, not present on the left side of eqn. (5.7), can be removed by setting $a_1b_2 = 0$ and $a_2b_1 = 0$. But if we do that, it is easy to see that we must also have either $a_1a_2 = 0$ or $b_1b_2 = 0$, in which case we will obtain only $|t\rangle_1|t\rangle_2$ or $|h\rangle_1|h\rangle_2$, respectively, these being just product states. So there is no way to retain only the terms $a_1a_2|h\rangle_1|h\rangle_2 + b_1b_2|t\rangle_1|t\rangle_2$ if we start out with a product state.

Another example of an entangled state is

$$\frac{1}{\sqrt{2}}\left(|h\rangle_1|t\rangle_2 + |t\rangle_1|h\rangle_2\right), \tag{5.10}$$

in which the coins are correlated in a different way: if 1 is heads, 2 is tails, and vice versa, and each occurs with the probability 1/2;. Many other such entangled states are possible, all of them being of the generic form

$$|\text{2coin state}\rangle = c_1|h\rangle_1|h\rangle_2 + c_2|t\rangle_1|t\rangle_2 + c_3|h\rangle_1|t\rangle_2$$
$$+ c_4|t\rangle_1|h\rangle_2.^* \tag{5.11}$$

We need to be very careful about this business of correlations. There is nothing inherently quantum about correlations *per se*; we encounter them in everyday life. If we roll a single die, and the number 1 comes up on top, then surely the number 6 is on the bottom. There is no

---

* This is an entangled states as long as the numbers c_1, c_2, c_3, and c_4 do not respectively equal the numbers a_1a_2, b_1b_2, a_1b_2, and a_2b_1, as otherwise the state of eqn. (5.11) will be just a product state as in eqn. (5.9).

mystery here; the correlations are built into the construction of dice. But suppose we have a *pair* of dice and discover that when we roll them, they *both* come up with the same number. Perhaps the number is different for each roll, but nevertheless, the same number comes up for each die on a given roll. Or, going back to coins, suppose that both come up heads or both come up tails after each flip. The observed correlations in these cases should make us suspicious. How can two independently flipped coins or rolled dice come up with identical readings?

When we flip real coins or roll real dice, these objects must obey the deterministic laws of classical physics, which predict that under identical initial conditions and with identical forces applied, identical results will follow. But it is impractical for identical initial conditions and forces to be applied for every coin-toss. That is why we would not expect the two coins to land the same way every time they are flipped. But, as a matter of *principle* within classical physics, it is possible to deterministically generate identical readings for two flipped coins. This means that under the conditions that prepare, say, the state $|h\rangle_1|h\rangle_2$, both coins will always come up heads, but the observed correlations are entirely classical, built in from the begining.

Another point that should be made is that correlations between two or more objects can be maintained over great distances. As a simple example, suppose we have a pair of gloves. In secret, each is placed in a separate but identical box, and one box is given to Alice and the other to Bob (Alice and Bob are the usual names given to the participants in two-party information protocols). They separate by some great distance. Perhaps Alice goes to Mars and Bob stays on Earth. If Alice then opens her box and finds a left-handed glove, she immediately knows that Bob has the right-handed glove in his box. Of course, if Bob opens his box and finds the right-handed glove, he knows that Alice has the left-handed glove. There is no mystery here. The correlations are built in from the beginning, just as they in principle could be for the coin states $|h\rangle_1|h\rangle_2$, $|t\rangle_1|t\rangle_2$, $|t\rangle_1|h\rangle_2$, and $|h\rangle_1|t\rangle_2$. None of this involves entanglement, nor would we expect entanglement for classical coins or gloves, and hence there is nothing inherently quantum here. In contrast, for quantum mechanical objects, entangled states are *superpositions* of correlated states, and thus we might think that such states may have other kinds of correlations that are strictly quantum. In fact, they do.

5.2 A Few Remarks about State Reduction and Projection

Before proceeding, it would be wise for us to make a few remarks about the matter of state reduction and its consequences when entangled states are involved. But first, we remind you of what happens for a single particle in a superposition state. If (again assuming that coins behave as quantum particles) we have the superposition state

$$|\psi\rangle = \frac{1}{\sqrt{2}}(|h\rangle + |t\rangle), \tag{5.12}$$

then a measurement of the state of the coin that can discriminate between heads and tails collapses the entire state $|\psi\rangle$ onto either $|h\rangle$ or $|t\rangle$. We represent these possible collapses as

$$|\psi\rangle \xrightarrow{\text{measuremnent}} |h\rangle, \text{ or} |\psi\rangle \xrightarrow{\text{measuremnent}} |t\rangle. \tag{5.13}$$

Notice that the factor $1/\sqrt{2}$—the probability amplitude for each of the states $|h\rangle$ and $|t\rangle$ in the superposition state $|\psi\rangle$—disappears along with the symbol for the state *not* detected. This simply means that there is now no uncertainty about the state of the coin. It is important to remember that a measurement does not reveal a pre-existing attribute of the coin. Rather, a measurement forces the particle to "take a stand", so to speak, such that the attribute being measured takes a definite value. That is the essence of eqn. (5.13). For real quantum particles such as photons, the particle itself may be destroyed in the process of measurement. Certainly, photodetectors of the type mentioned previously can destroy (absorb) the photons.

Now consider the entangled state

$$|\varphi\rangle = \frac{1}{\sqrt{2}}(|h\rangle_1|h\rangle_2 + |t\rangle_1|t\rangle_2). \tag{5.14}$$

We imagine that measurements are performed only on coin 1. If we find coin 1 in the state $|h\rangle_1$, then the two-particle state will have undergone the collapse,

$$|\varphi\rangle \xrightarrow[\text{on coin 1}]{\text{measuremnent}} |h\rangle_1|h\rangle_2. \tag{5.15}$$

No measurement was made on coin 2, but it will be found in the state $|h\rangle_2$ if a subsequent measurement is performed on it. This is because it is tightly correlated with coin 1. We say that coin 2 has been *projected* into the state $|h\rangle_2$ by the measurement performed on coin 1. Generally, as we have already measured its state, we shall not be further concerned with coin 1, so an alternative way of symbolizing the collapse and projection is to write

$$|\varphi\rangle \xrightarrow[\text{heads}]{\text{coin 1 detected to be}} |h\rangle_2. \qquad (5.16)$$

On the other hand, if coin 1 is found to be in the state $|t\rangle_1$, then coin 2 will be projected into the state $|t\rangle_2$, or symbolically

$$|\varphi\rangle \xrightarrow[\text{tails}]{\text{coin 1 detected to be}} |t\rangle_2. \qquad (5.17)$$

We shall use these ideas and this notation in a later section of the present chapter. The reduction of the state of one particle to a definite state as the result of the outcome of the measurement on another particle, with which it is correlated, is a feature of quantum mechanics.

As a preview of the consequences of entanglement, suppose the state above, $(|h\rangle_1|h\rangle_2 + |t\rangle_1|t\rangle_2)/\sqrt{2}$, is prepared and coin 1 is placed in a box and given to Alice, while coin 2 is also placed in a box but given to Bob. Alice and Bob separate as indicated in Figure 5.1.

Alice   Bob

Coin 1   Coin 2

Figure 5.1 Two coins are prepared in an entangled state and sent to distant observers Alice and Bob.

Now, if Alice looks and finds her coin to be heads, she knows that Bob will find his coin to be heads as well (we can switch the roles of Alice and Bob with no change in the argument that follows). The same will be true if she finds tails for her coin, as she will know that Bob's will also be tails. If this seems just like the case of product states of the coins, well, it really only *seems* this way, as there is a big difference. The difference lies in the fact that our prepared state is a *superposition* of $|h\rangle_1|h\rangle_2$ and $|t\rangle_1|t\rangle_2$, which in turn means that states of both coins together are objectively indefinite: it is not the case that the coins would be not $|h\rangle_1|h\rangle_2$ *or* $|t\rangle_1|t\rangle_2$ where observation by Alice and Bob merely reveals which. Rather, if Alice finds her coin to be heads, which she randomly will in 50% of the time that such states are prepared, Bob's will be heads too. The same reasoning holds if she gets tails. But here is the important point: As we just said, Alice's coin is *randomly* found to be either heads or tails (50% of the time heads and 50% of the time tails)—the same randomness that is at work in our earlier discussions on the superposition principle. This means that if Alice and Bob are separated by a great distance, then it *seems* that there must be some kind of rapid action-at-a-distance at work here: If Alice gets heads for coin 1, coin 2 seems to instantly "know", somehow, that it should also be heads, even though Bob can be very far away from Alice. Because of the element of randomness in the result of what state Alice finds her coin to be in, and because the coins (which we are pretending are quantum coins) can be vastly separated, the correlations arising in this case go beyond those that can be explained by any classical model. Real coins, of course, can never be in entangled states. But subatomic particles can be put into entangled states which reveal these strange correlation effects—these effects being the essence of entanglement. As we shall see, the effects of entanglement are even stranger than anything we have discussed up to this point.

5.3 Spooky Actions at a Distance: The EPR Argument

Is the moon still there when nobody looks?
ALBERT EINSTEIN

Albert Einstein was highly skeptical of the ontological aspects of quantum mechanics (that physical attributes of a system can be objectively indefinite), and by the fact it was a non-local theory as enunciated by the Copenhagenists, though neither party appreciated the full extent of the non-locality[†]. He had no doubt that the theory was successful in that it made correct predictions for the energy levels of atoms, and for atomic spectra, the structure of the atomic nucleus,

---

[†] A quick definition of "non-locality": It is the notion that distance objects cannot influence one another in a time faster than a light-signal can be sent between them. Such faster-than-light communications would appear to violate the theory of relativity.

and so on. However, Einstein felt that the theory was incomplete. It could not, after all, offer a good physical picture of what it means for, say, an electron to be in a superposition state of the type discussed in Chapter 2 in connection with electron interference experiments. An electron, Einstein would say, is a particle of a certain (albeit small) mass, and as such, it should have a definite location and a definite velocity (or momentum) at all times as it passes from the source to the detection screen. We may not know the values of these quantities, but they must surely exist, so Einstein would have said. This is in sharp contrast to the Copenhagen interpretation, which asserts that not only do we know nothing about the values of position or velocity, but that they have no objective existence until some intervening measurement causes the collapse of the wave function (or state vector) onto a definite value. Einstein, as well as many other physicists—such as Erwin Schrödinger, one of the founders of modern quantum mechanics— was repelled by this notion of wave-function collapse. Later in life Einstein would often ask colleagues questions such as "Is the moon still there when nobody looks?" The position and velocity of an electron, and of the Moon, should have definite values, even if unknown. These assumed definite values are also known as "elements of reality", to use a phrase from a famous paper entitled *Can Quantum-Mechanical Description of Physical Reality Be Considered Complete?*, published in the *Physical Review* in 1935 by Einstein, Boris Podolsky, and Nathan Rosen, and often referred to as the EPR paper. Our use here of the word "reality" is limited to the question of whether or not physical attributes of particles described by quantum mechanics can be objectively definite, that they exist prior to measurements. The assumption that they do exist is the assumption of "realism" in this restricted sense.

As disturbing as the lack of reality surely was to him, Einstein realized that quantum theory possessed an even more unsettling feature—namely, *non-locality*, or "spooky-actions-at-a-distance". Locality, sometimes called *Einstein locality*, is the premise that "signals", or more generally, "influences", cannot travel faster than the speed of light, or that distant objects cannot have instantaneous influences on each other. If the world, or at least the quantum world, is *non-local*, influences can travel faster than the speed of light, or may even be instantaneous over great distances. It *does not* mean that one can use those influences to signal, or communicate, at speeds faster than light. That would violate the principle of causality, meaning that it would be possible to find a uniformly moving frame of reference (one moving at constant speed in a constant direction) in which events could precede their causes, in conflict with the special theory of relativity. Such concerns were of vital importance to Einstein. After all, it was he who developed the special theory of relativity in 1905, out of which emerged the property of the speed of light as being the universal speed limit: nothing carrying energy and/or information can travel faster than the speed of light.

As already discussed in Chapter 4, superluminal effects are allowed (such as the speed of the rotating laser beam projected onto a distant screen), but they cannot be used to communicate. However, gravity and electromagnetism—the only fundamental interactions of nature known in 1905—must somehow convey real information in the form of forces, sometimes over vast distances. That electromagnetism can do so only in a local fashion (no faster than the speed of light) was shown to be the case in Einstein's 1905 paper "On the electrodynamics of moving bodies"—the paper that introduced the special theory of relativity. But the theory of gravitation at the time consisted of the simple force law derived by Isaac Newton some 300 years earlier: For two bodies of mass m_1 and m_2, separated by a distance d, there is an attractive force between them of magnitude $F = Gm_1m_2/d^2$, where G is the universal gravitation constant, which provides a measure of the intrinsic strength of the force. This is an action-at-a-distance force law, meaning that a disturbance in, say, the position of one mass, is instantly felt by the other. This, in turn, means that the law is *non-local* in the sense that distant events seem to have instantaneous effects over large distances. Newton's law of gravitation works well for relatively "small" masses (the Sun and the planets have small masses in this sense) with low velocities, and is in good agreement with most solar-system observations. However, it is not in agreement with the observed motion of the planet Mercury, which, being the planet closest to the Sun, feels the Sun's gravitational field the strongest. There was bound to be a failure of this law because it violates Einsteinian locality.[‡]

[‡] In 1915 Einstein presented a new theory of gravitation called the *general theory of relativity*, which, among other things, is a theory in which gravitational effects are local—that is, propagate at the speed of light. The new theory, in which gravity is a manifestation of curved spacetime, explained the observed motion of Mercury and predicted other effects such as the bending of light beams by gravitational fields and the probable existence of black holes. But this is not the place to elaborate on the theory of relativity. Rather, our intention is to highlight the importance of locality in Einstein's views on the nature of physical reality.

From the very beginning, Einstein could see a problem with non-locality in quantum mechanics, and discussed the issue at the 1927 Solvay Conference. If you think back to the electron interference experiment, recall that prior to the time that an electron lands on the screen and is detected, the wave function of a single electron is spread out (though not uniformly) over the screen. According to Copenhagen, the detection of an electron at one point on the screen instantaneously collapses the wave function to that point, such that there is now zero probability that the electron can be detected at some other point. This is a non-local, action-at-a-distance, effect that occurs through the assumption that the wave function is a complete mathematical description of the state of the electron. On the other hand, if the electron had a position at all times, its detection would merely reveal where it was, and that it was not somewhere else. But the idea that a particle passing through a double-slit apparatus has a definite location runs into trouble for reasons having to do with the formation of the interference pattern, as discussed in Chapter 2.

According to standard quantum theory, the microworld is neither local nor realistic in the senses just described. There is a class of theories, known as *realistic local hidden variable theories*, that are alternatives to quantum mechanics that attempt to address the supposed problems of reality and non-locality of the standard quantum theory. It is important to understand that these theories are not alternative interpretations of quantum mechanics but are really different theories. There are, in fact, two types of hidden variable theories: local and non-local. The non-local theories, of the type proposed by David Bohm in 1952, appear to solve the problem of reality but retain the feature of non-locality, the feature of quantum mechanics that bothered Einstein even more than the problem of reality.[§] On the other hand, local hidden variable theories, in certain cases, make predictions that differ from those of quantum mechanics, and these predictions can be subjected to experiments.

Einstein thought that the world must be local, that there could be no action-at-a-distance phenomena in nature, and that the attributes of particles must be real. The case for these ideas was laid out in his paper co-authored with Podolsky and Rosen (the EPR paper) of 1935,

---

[§] Einstein did not much care for Bohm's theory, saying that Bohm "got his results cheap." The Bohm theory does have its supporters—one, for example, being the philosopher David Z. Albert of Columbia University.

in which the conundrums associated with reality and locality were systematically presented for the first time. EPR provided the following very reasonable definition of an element of reality:

> If, without in any way disturbing a system, we can predict with certainty (i.e. with probability equal to unity) the value of a physical quantity, then there exists an element of physical reality corresponding to the physical quantity.

With a well-chosen example, EPR showed that this definition of elements of reality, and the requirement of locality as mentioned above, is problematic in quantum mechanics. We will shortly describe a convenient version of their example, but first we must take a detour in order to discuss the polarization states of photons.

5.4 Interlude: Polarization States of Photons

Recall from Chapter 2 that light can be polarized in any direction perpendicular to its direction of propagation. If light is shone onto a polarizing filter, only light polarized in one direction is allowed through. If the incident beam is polarized along the direction of a Polarizing filter axis, all of it will pass through. If polarized at a right angle to the filter axis, the light will be blocked.

The polarization of a single photon can be operationally defined by whether or not it passes through a polarizing filter. If it does, then it is polarized along the direction of the filter axis. If not, it is polarized along a direction perpendicular to that axis. Suppose we take the axis to be oriented in such a way that light that passes though the filter is vertically polarized, and light that is blocked by it is horizontally polarized. If a single photon is horizontally polarized, we denote its state as $|H\rangle$ (H is for horizontal polarization and not to be confused with h for heads) while if polarized vertically, we denote it as $|V\rangle$. On the other hand, photons could be polarized along the directions $\pm 45°$ to the H/V polarization directions. Photons with these polarizations are in states we denote as $|+\rangle$ and $|-\rangle$.** But these sets of states are not all independent of each other. For example, by looking

---

** These are shorthand for $|+45°\rangle$ and $|-45°\rangle$, respectively.

at Fig. 2.9 one can see that the $|+\rangle$ state is composed of equal amounts of $|H\rangle$ states and of the $|V\rangle$ states—that is, that

$$|+\rangle = \frac{1}{\sqrt{2}}[|H\rangle + |V\rangle], \qquad (5.18)$$

and similarly that

$$|-\rangle = \frac{1}{\sqrt{2}}[|H\rangle - |V\rangle]. \qquad (5.19)$$

The $1/\sqrt{2}$ factor is the usual normalization factor that occurs for "balanced" superpositions of two states. Note that by adding the $|+\rangle$ and $|-\rangle$[††] states we find that

$$|H\rangle = \frac{1}{\sqrt{2}}[|+\rangle + |-\rangle], \qquad (5.20)$$

whereas subtracting we obtain

$$|V\rangle = \frac{1}{\sqrt{2}}[|+\rangle - |-\rangle]. \qquad (5.21)$$

Thus, each set of polarization states, the $|\pm\rangle$ states or the $|H\rangle$ and $|V\rangle$ states, can be described as superpositions of the other. The fact that such superpositions accurately describe the quantum behavior of single photons means the underlying theory, quantum mechanics, is a *linear* theory.

In what follows we will need a way to analyze the polarization states of the photons. This can be done most simply with the use of calcite crystals. A calcite crystal has the property that a light-beam passing through it will be split into two beams along paths that depend on polarization—a consequence of the double-refraction property of calcite crystals, though the details of the double refraction process are unimportant here. If we place a crystal in a beam of unpolarized light, the light will emerge from it in two parallel beams of perpendicular polarization, as indicated in Figure 5.2(a), where we assume that the crystal is oriented to produce a horizontal and vertically polarized light beam. We label such a crystal an H/V

[††] When we add or subtract the $|+\rangle$ and $|-\rangle$ states, we are really adding or subtracting the numbers that multiply the states $|H\rangle$ and $|V\rangle$, and vice versa, to obtain the former, as in eqns. (5.20) and (5.21).

a)

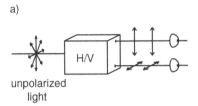

unpolarized
light

b)

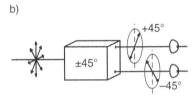

c)

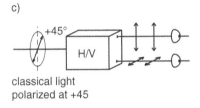

classical light
polarized at +45

d)

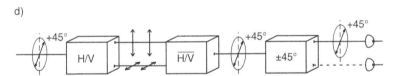

Figure 5.2. (a) Unpolarized light enters a calcite crystal (labeled H/V) oriented to produce horizontally and vertically polarized light. Light of these polarizations emerges in two separate beams. (b) Unpolarized light enters a calcite crystal oriented to produce $\pm 45°$ polarized light. (c) Light of polarization $+45°$ enters an H/V crystal and is split into beams of horizontal and vertically polarized light. (d) The horizontally and vertically polarized light generated from a beam of $+45°$ polarized light is recombined by a reversed calcite crystal, $\overline{H/V}$ to regenerate light of the original polarization, confirmed by the use of a $\pm 45°$ crystal. Only light of polarization $+45°$ emerges from this crystal.

crystal. On the other hand, if we rotate the crystal by 45°, unpolarized input light emerges in beams with polarization either $+45°$ or $-45°$, as indicated in Figure 5.2(b). A crystal producing light of this polarization we label a $\pm 45°$ crystal.

If light of polarization $+45°$ enters an H/V crystal it will emerge as horizontal and vertically polarized beams, just as does an initially unpolarized beam, as indicated in Figure 5.2(c). In what follows we will also require a *reversed* H/V crystal, which we will denote $\overline{\text{H/V}}$, which reunites H and V polarized beams to reproduce $+45°$ polarization. In Figure 5.2(d) we show what happens in the case where $+45°$ polarized light is split into H and V polarized beams, by an H/V crystal, and then recombined by an $\overline{\text{H/V}}$ crystal, the output of which is fed into a $\pm 45°$ crystal. All the light emerges along the $+45°$ beam out of the $\pm 45°$ crystal, thus confirming that the back-to-back crystals do indeed recombine the H and V polarized beams.

Let us now consider what happens at the level of a single photon. Suppose we perform a sequence of experiments where we inject, one at a time, photons in the $|+\rangle$ state into the H/V crystal, as in Figure 5.2(c). This state is a superposition of the $|H\rangle$ and $|V\rangle$ states as given in eqn. (5.18) above. In many runs of the experiment, about half will emerge in the H beam, and the other half in the V beam. But, as per the superposition principle, this *does not* mean that 50% of the incident $+45°$ photons actually *were* in the H or V states, and that the crystal plus detectors merely revealed which state. Rather, the photon's polarization with respect to H and V is objectively indefinite (indeterminate), and it is the detectors placed at the output channels of the crystal that causes the collapse of the quantum state of each photon randomly onto H or V.

Were this not the case, it would be hard to explain what happens in experiments where we put a reversed calcite crystal, $\overline{\text{H/V}}$, back to back with the first one, followed by a $\pm 45°$ calcite crystal, as indicated in Figure 5.3. Assuming photons prepared in the state $|+\rangle$ enter the H/V crystal one at a time in each of our experiments, we first consider the case when the horizontally polarized output beam of the H/V crystal is blocked, as in Figure 5.3(a). Then, only vertically polarized photons enter the $\overline{\text{H/V}}$ crystal, which produces a beam containing either a $+45°$ or $-45°$ photon, as will be verified using a $\pm 45°$ crystal followed by detectors in the output beams. That is because the vertically polarized photon state is a superposition of

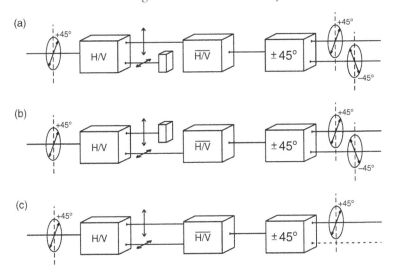

Figure 5.3. A series of experiments with two calcite crystals, one reversed, with an incident photon in the state $|+45°\rangle$. (a) The horizontally polarized beam is blocked, (b) the vertically polarized beam is blocked, (c) neither beam is blocked.

the $|\pm\rangle$ states as given in eqn. (5.21) above. If we instead block the upper beam, as shown in Figure 5.3(b), then only horizontally polarized photons enter the $\overline{H/V}$ crystal, and we have the same result as before: the production of either a $+45°$ or $-45°$ photon.

But now consider what happens if neither beam is blocked, as shown in Figure 5.3(c). In this case we have an informational ambiguity. We simply do not know, and cannot know, which path is taken to the second crystal by the input photon. So, following the rule that ambiguity leads to quantum interference, we should expect a result different from those of the previous two cases where we knew which path the photons had to take to the second crystal. And we do, in fact, obtain a different result. When the beams are recombined, only a $+45°$ polarization photon emerges, and that happens in *every run* of the experiment if the incoming photon has polarization $+45°$. If the incoming $+45°$ photon *also* had a definite polarization with respect to H and V, say H, then the second crystal should produce $-45°$ photons as well as $+45°$ photons. But that does not happen because the actual state of the photons between the crystals is still

$|+\rangle$. But suppose one took the view that a $+45°$ photon has an objectively definite polarization state with respect to H and V if it were in the states $|+\rangle$ *and* $|H\rangle$ simultaneously. Then, because $|H\rangle$ is a balanced superposition of the $|+\rangle$ and $|-\rangle$ states as given in eqn. (5.20) above, we would expect to get, out of a series of a large number of runs, with the input photon prepared in the state $|+\rangle$, about 50% of them to emerge as $+45°$ polarized, and the others to emerge as $-45°$ polarized photons. There would be no difference at all between the cases where one of the beams is blocked and where neither is blocked. But where neither beam is blocked, only $+45°$ polarized photons emerge.

Note that this is the same *kind* of experiment discussed a few paragraphs above, with classical light beams, for which there is no conundrum. It all makes sense from a classical point of view. A light-beam containing many photons with polarization $+45°$ is split into H and V components, each with a large number of photons, which are then recombined to produce the original beam—an unsurprising result. But at the level of a single photon, the photon emerges with its initial $+45°$ polarization—a result difficult to reconcile with the classical viewpoint. After all, we have only one photon.

However, we can explain our result as being due to quantum <u>interference</u> as follows: Combining of the $|H\rangle$ and $|V\rangle$ beams by the H/V crystal causes the cancellation of the $|-\rangle$ state, as we can see by using eqns. (5.18)–(5.21), to obtain

$$
\begin{aligned}
|H\rangle + |V\rangle &= \frac{1}{\sqrt{2}}[|+\rangle + |-\rangle] + \frac{1}{\sqrt{2}}[|+\rangle - |-\rangle], \\
&= \sqrt{2}|+\rangle, \\
\Rightarrow |+\rangle &= \frac{1}{\sqrt{2}}[|H\rangle + |V\rangle].
\end{aligned}
\tag{5.22}
$$

The point of all of this is that superpositions of quantum states are not the same thing as mixtures of quantum states.[‡‡] Mixtures lack the quantum coherence that allows quantum superpositions to produce results like that of eqn. (5.22). Mixtures are essentially classical in nature.

[‡‡] Recall that we discussed mixtures in Chapter 2. There are no probability *amplitudes* in mixtures, only probabilities, so there is no possibility for quantum interference.

5.5 Back to EPR

We now discuss a convenient form of the conundrum presented by EPR in the form of a thought experiment. We consider a state describing two photons, labeled A and B, entangled through their polarizations. We take $|H\rangle_A$ and $|V\rangle_A$ to mean states where photon A is horizontally and vertically polarized, respectively, and $|H\rangle_B$ and $|V\rangle_B$ mean the same for photon B. Our entangled state of the two photons we choose as

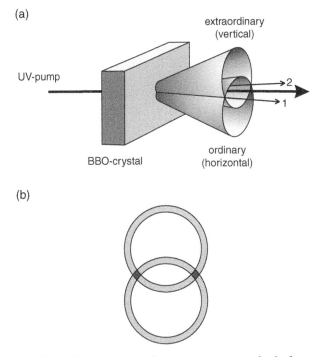

Figure 5.4. Type II spontaneous down-conversion method of generating polarization entangled photons. (a) Photons are emitted from the non-linear crystal along two cones, the upper cone containing the *e*-rays where the photons are vertically polarized, the lower cone containing the *o*-rays where the photons are horizontally polarized. (b) At the two intersection points of the cones, there is ambiguity as to the polarization of the emergent photons.

$$|\psi\rangle = \frac{1}{\sqrt{2}}[|H\rangle_A|V\rangle_B + |V\rangle_A|H\rangle_B]. \qquad (5.23)$$

Such states can be produced in the laboratory, though the details of how this is done are not of paramount interest here.[§§] (See Figure 5.4.) We will assume that such states are available in the laboratory.

Suppose we arrange for photons A and B to move in opposite directions from the source that produced them (mirrors may be required), and toward distant observers Alice and Bob, each equipped with calcite crystals and detectors, as pictured in Figure 5.5. Alice and Bob are assumed to be widely separated spatially, such that the measurements they perform do not have time to influence what happens at the other's distant location by any signal traveling at the speed of light. Suppose Alice measures the H/V polarization of photon A. She has a 50:50 chance of getting either H or V. Let us say she gets H. Then, because of the correlations embedded in the state $|\psi\rangle$, $|\psi\rangle$ *instantly* reduces to $|V\rangle_B$:

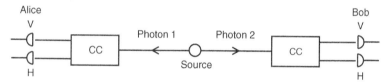

Figure 5.5. Two polarization entangled photons are prepared and directed to distant observers Alice and Bob. Alice gets photon 1 and Bob gets photon 2. They perform polarization measurements on their respective photons. CC denotes calcite crystal.

[§§] Such a state can be generated by a process called *type II down-conversion*, which is illustrated in Figure 5.4. It is similar to type I down-conversion discussed in Chapter 4, where photons are emitted along two cones. But here, the two cones—one containing what the opticians call the ordinary rays (*o*-rays) and the other containing the extraordinary rays (*e*-rays)—are not concentric, nor are the polarizations of the photons in two cones the same. The photons in *o*-rays and *e*-rays, the two cones, are polarized at 90° to each other. Note that the cones intersect. Let us say that those of the *o*-ray are horizontally polarized and those of the *e*-ray vertically polarized. If the experimenter uses a two-hole "mask" over the output of the crystal, with the holes placed over the intersection points, all the photons will be blocked except those pairs of photons coming from the two intersection points. This creates an ambiguity in regard to the photons coming from these points: Is a photon coming from the *o*-ray or the *e*-ray? Once again, ambiguity, the lack of information, is characteristic of quantum superposition, so in this case we have to superpose the product states $|H\rangle_1|V\rangle_2$ and $|V\rangle_1|H\rangle_2$ to get $|\psi\rangle$ above. The subscripts 1 and 2 label the beams coming from the intersection points of the cones.

$$|\psi\rangle \xrightarrow{\text{Alice finds photon } A \text{ to be } H} |V\rangle_{\text{B}}.$$

The instantaneous reduction of the state of the distant photon B as a result of Alice detecting photon A to be H polarized invokes the locality problem that so worried Einstein. The crucial point of EPR is that the only measurement performed is on photon A, and nothing has interacted with photon B. Photon B must have a well-defined polarization V—a prediction that would be confirmed if Bob *were* to perform an H/V measurement (but he does not). Thus, the predicted polarization V of photon B is an element of reality in the sense of EPR. If instead Bob were to perform a $\pm45°$ measurement, he would get $+45°$ or $-45°$ on a 50:50 basis, but we assume that he does not do this measurement either.

On the other hand, suppose Alice performs a $\pm45°$ polarization measurement on photon A. To see explicitly what happens, let us first use eqns. (5.20) and (5.21) to put the state of her photon in terms of its $\pm45°$ polarization states. By simple substitution we will find that $|\psi\rangle$ can be written as

$$|\psi\rangle = \frac{1}{\sqrt{2}}\left[\frac{1}{\sqrt{2}}\left(|+\rangle_{\text{A}}+|-\rangle_{\text{A}}\right)|V\rangle_{\text{B}}+\frac{1}{\sqrt{2}}\left(|+\rangle_{\text{A}}-|-\rangle_{\text{A}}\right)|H\rangle_{\text{B}}\right],$$
$$= \frac{1}{2}\left[|+\rangle_{\text{A}}\left(|V\rangle_{\text{B}}+|H\rangle_{\text{B}}\right) + |-\rangle_{\text{A}}\left(|V\rangle_{\text{B}}-|H\rangle_{\text{B}}\right)\right]. \tag{5.24}$$

If Alice finds photon A to be in, say, the $+45°$ state, we have the instantaneous reduction

$$|\psi\rangle \xrightarrow{\text{Alice finds photon } A \text{ to be}+} \frac{1}{\sqrt{2}}\left(|V\rangle_{\text{B}}+|H\rangle_{\text{B}}\right) = |+\rangle_{\text{B}}, \tag{5.25}$$

from which we see that Bob's photon *must* be in the state $|+\rangle_{\text{B}}$. Thus, the predicted $+45°$ polarization for photon B must also be an element of reality.

But there is a problem. The predicted elements of reality ought to be definite—that is, ought to exist whether or not anyone carries out a measurement to check. But according to quantum mechanics, the kind of measurement performed on photon A, and the results of that measurement, instantly dictates the state of the photon B. If Alice gets H, photon B must be in the state $|V\rangle_{\text{B}}$, while if she gets $+45°$, photon B is in the state $|+\rangle_{\text{B}}$. These predictions about photon B constitute elements of reality in the sense that, in the first instance, if Bob performs an

H/V polarization measurement on his photon, he will surely get V, while in the second instance, should he perform a $\pm 45°$ measurement he will surely get $+45°$. But these states are not independent of each other through the superposition principle—that is, because

$$|+\rangle_B = \frac{1}{\sqrt{2}}[|H\rangle_B + |V\rangle_B],\qquad(5.26)$$

and

$$|V\rangle_B = \frac{1}{\sqrt{2}}[|+\rangle_B - |-\rangle_B].\qquad(5.27)$$

Eqn. (5.26) implies that the H/V polarization of photon B is indeterminate, while eqn. (5.27) implies that its $\pm 45°$ polarization is indeterminate. Thus, neither the V nor $+45°$ polarizations of photon B can be elements of reality. The conundrum, brought to light by EPR, that these contradictory results arise from measurements on photon A having an instant effect on the state of B, is often called the *EPR paradox*. This "paradox" led EPR to conclude that quantum mechanics must be an incomplete theory.

Naturally, the EPR paper (which came as "a bolt from the blue", according to one of Bohr's associates) caused a stir within the atomic-physics community. Bohr himself temporarily dropped all of his other work (mostly in nuclear physics at the time) to deal with the EPR conundrum, which was, after all, a direct attack on the Copenhagen interpretation. Leon Rosenfeld, a close associate of Bohr, said that "...we had to clear up such a misunderstanding at once." Notice his automatic assumption that it was EPR who had displayed a "misunderstanding", with no admission of the possibility that they had raised a profound issue that needed to be explored. (The attitude that the Copenhagen school was unquestionably correct in all matters regarding the interpretation of quantum mechanics will be discussed again in Chapter 8.) Bohr's response came in a paper with the same title as EPR's, and was also published in *Physical Review* in 1935. In terms of the photon-polarization-state version of the EPR paradox as given above, Bohr conceded that a measurement of photon A would cause no disturbance on photon B. Then he goes on to say: "But even at this stage there is essentially the question of an *influence on the very conditions which define the possible types*

of predictions regarding the future behavior of the system" (Bohr's emphasis).
Another way to state Bohr's response is as follows:

> We must consider the experimental situation as a whole
> and not ask questions about the separate parts; that is
> to say we must not ask questions about the properties
> of the individual particles independent of the kinds of
> measurements performed on them.

In other words, we cannot perform a measurement on one part of
a quantum mechanical system and leave the state of the other part
undisturbed even though there has been no disturbance (measurement) on that part. That the state of one part of a system is determined automatically, as if by telepathy, as a result of a measurement
on another (distant) part without any obvious causal (local) mech-
anism was precisely the "spooky action at a distance" that caused EPR
to conclude that quantum mechanics was an incomplete theory.
It was, and perhaps still is, widely thought by the physics community
that Bohr won the argument with EPR. But a close examination of
Bohr's response reveals that he merely restates the conundrum that
they exposed, as was pointed out by Howard Wiseman, of Griffith
University, in a recent paper in the journal *Contemporary Physics*. Bohr's
response is just a rephrasing of the EPR paradox! Most of Bohr's
response does not even address the conundrum of the EPR thought
experiment itself. Rather, it appears that Bohr's primary interest was
in showing, through the principle of complementarity, that quantum
mechanics was a *consistent* theory, though EPR never suggested otherwise. Their quarrel concerned its completeness.

The alert reader will have noted that Bohr's reply to Einstein
contains a line of reasoning that we have seen before. In fact, the
explanation of every quantum mystery invokes it in some way. It is,
after all, the essence of the Copenhagen interpretation. And that
interpretation fits fairly well into the philosophical school of thought
known as *logical positivism*, wherein one can only speak of measurable
quantities as having real existence. Recall Wittgenstein: "What can be
said at all can be said clearly; and whereof one cannot speak thereof
one must be silent." Indeed, Bohr said (in a letter to Aage Peterson,
one of his assistants):

> There is no quantum world. There is only an abstract
> physical description. It is wrong to think that the task of

physics is to find out how nature is. Physics is concerned with what one can say about nature.

Bohr's answer to the EPR paper never persuaded Einstein, who maintained to the end of his life that quantum mechanics is incomplete as exhibited by both the non-locality of the theory and the fact that it lacked the ability to assign definite values (elements of reality) to all physical quantities.

To complete the theory in Einstein's sense, one presumably needed to supplement standard quantum theory with so-called "local hidden variables". But to most physicists the whole issue appeared to be essentially a philosophical one, as for a long time (indeed, from 1935 right through the mid-1960s) it was not thought possible to decide one way or the other by experiment. Local hidden variable theories were not thought to predict results different from those of standard quantum mechanics. Thirty years after the publication of the EPR paper, however, the situation changed dramatically due to the work of John Bell, which we shall discuss in the next section.

We close this section by mentioning briefly that in 1935 Schrödinger also wrote a paper in response to EPR, wherein he pointed out that the essential feature of the system in their thought experiment, and of quantum mechanics in general, was "entanglement". This was the first use of the term "entanglement" in physics in this context, referring to the non-separability of two-particle quantum states. We will return to Schrödinger's 1935 paper in Chapter 7.

5.6 Bell's Theorem

John Bell—an Irish physicist who spent most of his career at the European Center for Nuclear Research (CERN)—when working in the early 1960s thought little of Bohr's response to EPR. In fact, he thought Einstein's various writings on the topic, as on all other topics, exceptionally clear, whereas he thought Bohr's writings exceptionally obscure. Many who have read Bohr's writings on complementarity and other philosophical aspects of quantum theory have noted opacity rather than clarity. In spite of this, Bohr and his followers apparently had convinced themselves that the matter was settled by Bohr's response to EPR, and that there was simply nothing more to be said. Besides, in 1935 there seemed to be no experimental consequences to the EPR conundrum, and, as we said, physicists thought the argument over Copenhagen versus local realism entirely philosophical. But in 1964 Bell showed that philosophy could be subjected to experimental tests and that Bohr had not settled all possible questions about the meaning of quantum mechanics—or at least, he had not *obviously* settled them.

In the mid 1960s, Bell, assuming that all physically relevant quantities did possess definite values (that is, elements of reality), as determined by hidden variables, showed that local hidden variable theories do make predictions different than those of standard quantum mechanics in some cases. This led him to formulate a way to distinguish between the two kinds of theories in the laboratory. It turned out that local hidden variable theories make a prediction on the allowed values of a combination of correlations determined statistically from experimental data. The allowed values are bounded mathematically—that is, they should not exceed a certain numerical value. This mathematical relation is known as *Bell's inequality*. Local hidden variable theories satisfy Bell's inequality; but quantum mechanics can *violate* Bell's inequality.

Starting in the late 1960s there began a series of experiments designed to discriminate between the two theories. The experiments were optical and involved pairs of polarized photons, and variations on these experiments continue to this day. In the early days, most of the physicists involved—such as John Clauser, then at Lawrence Livermore Laboratory, who spearheaded the initial efforts toward laboratory tests—thought that local hidden variable theories would be supported. They were wrong: Bell's inequality was found to be violated, thus refuting local hidden variable theories. These early experiments used atomic transitions for sources of entangled light. The most spectacular demonstration of the violation of Bell's inequality using such sources was in an experiment performed by A. Aspect, P. Grangier, and G. Roger (the trio already mentioned in Chapter 3 in connection with the single-photon interference experiment) in 1982. That experiment provided the most convincing evidence local hidden variable theory to that date. Starting in the late 1980s, Bell's inequality experiments have used type II down-conversion as a source of entangled photons. In an experiment performed in 2005 by J. B. Altepeter, E. Jeffrey, and P. Kwiat at the University of Illinois, the inequality was violated by more than 1,200 standard deviations—the standard deviation being a statistical measure of the spread of the data. In this experiment the spread is very narrow, and hence constitutes the largest violation of Bell's inequality reported to date, as far as we are aware.

In spite of this result, this, and all the Bell inequality violating experiments performed to date, are not perfect and possess loopholes through which local hidden variable theory explanations of the results can creep through. With respect to Bell-inequality experiments involving light-beams, there is only one loophole, and that has to do with photodetector efficiency. The efficiency of such devices is quite low, and because of this it is possible to explain the results of the experiments with a local, realistic theory. Attempts are under way to circumvent the detection loophole by using a different kind of detection scheme that is nearly 100% efficient. This requires a rather different set-up and even a different kind of quantum state than what was used in the experiment by Kwiat's group. We shall not describe these attempts further. We are not aware of any such loophole-free Bell inequality experiments performed as of this writing.

We will not discuss Bell's inequalities further, nor will we describe any of the above-mentioned experiments. Bell's inequalities are of a statistical nature, and many runs of an experiment are required in order to accumulate sufficient data. The violation of Bell's inequality, and how this falsifies local hidden variable theories, is rather subtle. Fortunately, a way has been proposed to falsify local hidden variable theories without the use of an inequality.

In fact, two ways of accomplishing that have been proposed and experimentally demonstrated. In 1989 Daniel Greenberger, of the City College of the The City University of New York, Michael Horne, of Stonehill College in Massachusetts, and Anton Zeilinger, of the University of Vienna, proposed a method that involved entangled states (known as GHZ states) of three or more entangled particles. Inspired by their argument, Lucien Hardy proposed an argument involving just two particles. Because of its simplicity, at least in a relative sense, we present here the Hardy argument as later improved by Thomas Jordan.

5.7 The Hardy–Jordan Argument: Bell's Theorem without Inequalities

The argument we present was proposed in 1993 by Lucien Hardy, at the time of St Patrick's College, Kildare, Ireland, and later refined in

1994 by Thomas Jordan of the University of Minnesota, Duluth. The state they use involves two photons in a polarization entangled state, similar to the types of states discussed above in connection with Bell's inequality, but also different in an important way.

Before introducing the state itself, we wish to again invoke the assistance of two experimenters, Alice and Bob, who are well separated as before, and who each have one photon of the prepared state of the two photons. As before, Alice's photon is labeled A and Bob's is labeled B. Alice and Bob can perform two kinds of polarization detection measurement on their photons: They can do H/V polarization measurements, or they can do $+45°$ polarization measurements. Now consider the following set of measurements and results.

1. If Alice performs an H/V measurement and gets V, then Bob will get $+45°$ if he does a $\pm 45°$ measurement.
2. If Bob performs an H/V measurement and gets V, then Alice will get $+45°$ if she does a $\pm 45°$ measurement.
3. If both Alice and Bob perform H/V measurements, they will sometimes both get V.
4. If Alice and Bob both perform $\pm 45°$ measurements they *never* both observe $+45°$.

We now show that local realism forces us to conclude that not all four conditions can be true. With regard to condition (1), Alice's H/V polarization measurement resulting in V predicts that Bob will get $+45°$ if he does a $\pm 45°$ polarization measurement, meaning that his photon's polarization $(+45°)$ is an element of reality in the sense of EPR, whether or not Bob actually performs his measurement. The requirement of locality demands that this element of reality be independent of what kind of measurement Alice performs. It also means that Bob's photon must have had the property of having $+45°$ polarization from the moment it was created. By the same reasoning, condition (2) shows that Alice's photon also has polarization $+45°$ as an element of reality, and must have had that property from the moment of its creation. So, *both* photons have, as elements of reality, polarization of $+45°$. The fact that if both Alice and Bob perform H/V polarization measurements and occasionally both get V polarizations of their photons, as specified in condition (3), suggests that if instead they *had* performed $\pm 45°$ polarization measurements, they should, according to our conclusions about the results of conditions (1) and (2), occasionally both obtain the result $+45°$ for their respective photons. But this is in contradiction to condition (4): they should *never* simultaneously observe $+45°$ polarization of their photons. To be clear: the contradiction occurs because the $+45°$ polarizations of the two photons were established as elements of reality by conditions (1) and (2). Local realism implies that Alice and Bob *should* find both their photons to occasionally have polarization $+45°$ if they perform $\pm45°$ polarization measurements. The conditions (1)–(4) are incompatible in a world where local realism holds. But as we shall show, quantum mechanics can be used to satisfy those conditions, provided that Alice and Bob possess photons in a particular kind of entangled state.

An entangled state of two photons that leads to the satisfaction of conditions (1)–(4) is the following:

$$|\psi\rangle = N\left[|H\rangle_A|H\rangle_B - \frac{1}{2}|+\rangle_A|+\rangle_B\right]. \qquad (5.28)$$

The factor N is a normalization factor whose value is of no importance to the demonstration. Note that this state is a superposition of product states defined by the two types of measurement: H/V polarization measurements and $\pm45°$ polarization measurements. However, the state can be rewritten in terms of the states associated with either of these measurement schemes, as needed, by using the relations in eqns. (5.18)–(5.21) above. So, we now proceed to show how this state can satisfy conditions (1)–(4) above:

(1) Here, Alice makes an H/V measurement and Bob makes a $\pm45°$ measurement. This means that we must express all parts of the state $|\psi\rangle$ involving Alice's photon in terms of the $|H\rangle_A$ and $|V\rangle_A$ states, and we must express all parts of the state involving Bob's photon in terms of the $|+\rangle_B$ and $|-\rangle_B$ states. So, using eqns. (5.18) and (5.21) we can rewrite our state of eqn. (5.28) as

$$|\psi\rangle = \frac{N}{\sqrt{2}}\left[|H\rangle_A(|+\rangle_B + |-\rangle_B) - \frac{1}{2}(|H\rangle_A + |V\rangle_A)|+\rangle_B\right]. \qquad (5.29)$$

We can collect some like terms and rewrite this as

$$|\psi\rangle = \frac{N}{\sqrt{2}}\left[\frac{1}{2}|H\rangle_A|+\rangle_B + |H\rangle_A|-\rangle_B - \frac{1}{2}|V\rangle_A|+\rangle_B\right]. \qquad (5.30)$$

Recall that products of the states for the two systems imply correlations in measurement outcomes. Here, specifically, we can see that if Alice performs an H/V measurement and obtains the result V, then, because of the correlations in the last term, the one with $|V\rangle_A|+\rangle_B$, Bob will indeed obtain the result $+45°$ if he performs a $\pm45°$ measurement. (Notice that if Alice gets the result H, Bob could get either $+45$ or $-45°$, with a greater likelihood of getting the latter.)

(2) Alice and Bob switch their measurements types—Alice performing $\pm45°$ measurements, and Bob performing H/V measurements. This now requires us to do the opposite of what we just did: in eqn. (5.28) we replace $|H\rangle_A$ and $|+\rangle_B$ using eqns. (5.18) and (5.20). Doing so will result in

$$|\psi\rangle = \frac{N}{\sqrt{2}}\left[\frac{1}{2}|+\rangle_A|H\rangle_B + |-\rangle_A|H\rangle_B - \frac{1}{2}|+\rangle_A|V\rangle_B\right], \qquad (5.31)$$

where it is evident, because of the last term, $|+\rangle_A|V\rangle_B$, that when Bob performs an H/V measurement on his photon and gets V, Alice will get $+45°$ if she performs a $\pm45°$ measurement on her photon.

(3) Both Alice and Bob perform H/V measurements. This means that the term $|+\rangle_A|+\rangle_B$ in our state of eqn. (5.28) must be expressed in terms of H and V states. Using eqn. (5.18) twice, we have

$$|\psi\rangle = N\left[|H\rangle_A|H\rangle_B - \frac{1}{4}(|H\rangle_A + |V\rangle_A)(|H\rangle_B + |V\rangle_B)\right], \qquad (5.32)$$

or, upon writing out all the products and collecting common terms, we have

$$|\psi\rangle = N\left[\frac{3}{4}|H\rangle_A|H\rangle_B - \frac{1}{4}|V\rangle_A|H\rangle_B - \frac{1}{4}|H\rangle_A|V\rangle_B - \frac{1}{2}|V\rangle_A|V\rangle_B\right]. \qquad (5.33)$$

We see, from the last term, the $|V\rangle_A|V\rangle_B$ term, that Alice and Bob will sometimes both get V. (They can get other results as well—for example, H–H, H–V, and V–H.)

(4) Because Alice and Bob are both to perform $\pm45°$ measurements, we must express the $|H\rangle_A|H\rangle_B$ term of our original state of eqn. (5.28) in terms of the $\pm45°$ states to obtain

$$|\psi\rangle = N\left[\frac{1}{2}(|+\rangle_A + |-\rangle_A)(|+\rangle_B + |-\rangle_B) - \frac{1}{2}|+\rangle_A|+\rangle_B\right]. \qquad (5.34)$$

If we multiply everything out, we find that the $|+\rangle_A|+\rangle_B$ term disappear altogether; that is, our state in terms of the $\pm 45°$ will look like this:

$$|\psi\rangle = \frac{N}{2}[|+\rangle_A|-\rangle_B + |-\rangle_A|+\rangle_B + |-\rangle_A|-\rangle_B]. \qquad (5.35)$$

Because the $|+\rangle_A|+\rangle_B$ terms cancel out and do not appear in the final expression of the state $|\psi\rangle$ in terms of the $\pm 45°$ states, there is no possibility of Alice and Bob ever detecting those states in their measurements. The non-detection of those results satisfies condition (4) above, whereas local realism cannot satisfy it. Quantum superposition and interference have conspired to remove this possibility—a feat not possible if local realism holds, as we have argued above. Thus, an experiment able to produce the state of eqn. (5.28), and able to perform the indicated measurements, can falsify local realism because of the inability to detect the $|+\rangle_A|+\rangle_B$ combination, as specified above.

The state that we have used, given in eqn. (5.28), is just one particular state of two photons that can be used to show a violation of local realism by the Hardy–Jordan argument.

Several experiments have, in fact, been performed to test the Hardy–Jordan proposal for demonstrating the non-local and non-realistic nature of quantum mechanics. The first was performed in 1995 by Leonard Mandel's group at the University of Rochester. A sketch of their experiment is given in Figure 5.6. Their light-source, as in all such experiments involving two polarization entangled photons, was a down-conversion process which produces a pair of photons of identical polarization. The photon polarization in one beam is rotated (by rotator R_0) so as to be perpendicular to the polarization of the photon in the other beam. The photons are then sent to a beam-splitter, which mixes the states and emits the photons in opposite directions. Because the photons striking the beam-splitter are not identical (they have different polarizations), the Hong–Ou–Mandel result does not apply, and there will be four possible outcomes: both photons are reflected, both transmitted, one reflected and the other transmitted, and the same with the photons interchanged. The experimenters consider only the first two cases wherein both detectors register a count. By using a subsequent polarization rotator in each beam, and the fact that the

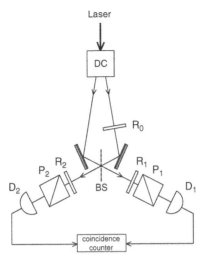

Figure 5.6 A sketch of the experiment of Torgerson, Branning, Monken, and Mandel. Pairs of identically polarized photons are produced be the down-conversion process. R_0, R_1, and R_2 rotate the polarizations of the photons in their respective beams. R_0 rotates the polarization of the photon it its beam by $90°$ so as to be perpendicular to the polarization of its partner photon in the other beam. The only runs of the experiment that are considered are those where both detectors are fired.

beam-splitter employed was not a 50:50 beam-splitter (meaning that the probabilities of reflection or transmission are different going in one direction as opposed to the other direction), it is possible to manipulate the state before detection to be one of the forms of the Hardy–Jordan state. The results of the experiments on the whole support quantum mechanics. We should emphasize that none of the experiments testing local hidden variable theories, or local realism in general, is perfect due to the lack of 100%-efficient photon detectors. Yet the experiments on the Hardy–Jordan state violated the predictions of local realistic theories by 45 standard deviations, and thus on the whole supports quantum mechanics.

5.8 What Should we Give Up? Locality, Realism, or Both?

One could ask: What is the fundamental "problem" with why quantum mechanics is at such odds with our classical reasoning in experiments like the Hardy–Jordan experiment just described? Is it Einsteinian locality, the assumption of realism (as that word is used in this context), or some combination of both? In 2003, Anthony Leggett, from whom we shall hear more in Chapter 7, proposed a test of *non-local* realistic theories—that is, alternatives to quantum mechanics that are non-local (that violate Einsteinian locality) but assume the existence of "elements of reality" in the sense of EPR. Leggett devised a mathematical inequality, similar to Bell's inequality, given in terms of certain experimentally observable quantities, which, if violated by experiments, falsifies non-local realistic alternatives to standard quantum mechanics. In 2007, two groups—one headed by Nicolas Gisin of the University of Geneva and Valerio Scarani of the National University of Singapore, and another headed by Zeilinger in Vienna—reported experimental results that have shown violations of Leggett's inequality and thus have falsified his non-local model.

These results suggest that it is not enough to give up the notion of Einsteinian locality in order to explain quantum experiments. Rather, one must also let go of realism—the "elements of reality" whose existence was taken to be intuitively obvious by EPR.

Bibliography

Albert D. Z., *Quantum Mechanics and Experience*, Harvard University Press, 1992.

Altepeter J. B., Branning E. R., and Kwiat P. G., "Phase-compensated ultrabright source of entangled photons", *Optics Express* 13 (2005), 8951.

Aspect A., "Bell's theorem: the naïve view of an experimentalist", in *Quantum [Un]speakables—From Bell to Quantum Information,* ed. A. Bertlemann and A. Zeilinger, Springer, 2002.

Aspect A., Grangier P., and Roger G., "Experimental realization of Einstein–Podolsky–Rosen–Bohm Gedanjenexperiment: A new violation of Bell's inqualities", *Physical Review Letters* 49 (1982), 91.

Bell J. S., "On the Einstein–Podolsky–Rosen Paradox", *Physics* 1 (1964), 195.

Bell J. S., *Speakable and Unspeakable in Quantum Mechanics*, Cambridge University Press, 1987.

Bohr N., "Can quantum-mechanical description of physical reality be considered complete?", *Physical Review* 48 (1935), 696.

Clauser J. F., and Shimony A., "Bell's theorem: experimental tests and implications", *Reports on Progress in Physics* 41 (1978), 1881.

Einstein A., Podolsky B., and Rosen N., "Can quantum-mechanical description of physical reality be considered complete?", *Physical Review* 47 (1935), 777.

Freire O., Jr., "Philosophy enters the optics laboratory: Bell's theorem and its first experimental tests (1965–1982)", *Studies in the History and Philosophy of Modern Physics* 37 (2006), 577.

Genovese M., "Research on hidden variable theories: A review of recent progress", *Physics Reports* 413 (2005), 319.

Greenberger D. M., Horne M. A., and Zeilinger A., in *Bell's Theorem, Quantum Theory, and Conceptions of the Universe*, ed. M. Kafatos, Kluwer Academic, 1989.

Greenberger D. M., Horne M. A., Shimony A., and Zeilinger A., "Bell's theorem without inequalities, *American Journal of Physics* 58 (1990), 1131.

Hardy L., "Non-locality for two particles without inequalities for almost all entangled states", *Physical Review Letters* 71 (1993), 1665.

Jordan T. F., "Testing Einstein–Podolsky–Rosen assumptions without inequalities with two photons or particles with spin $\frac{1}{2}$", *Physical Review A* 50 (1994), 62.

Mermin N. D., "What's wrong with these elements of reality?", *Physics Today* 43, 6 (1990), 9.

Pan J.-W., Bouwmeester D., Daiell M., Weindofer H., and Zeilinger A., "Experimental test of quantum mechanical non-locality in three-photon Greenberger–Horne–Zeilinger entanglement", *Nature* 403 (2000), 515.

Torgerson J., Branning D., and Mandel L., "A method for demonstrating violation of local realism with a two-photon down-converter without use of Bell inequalities", *Applied Physics B* 60 (1995), 267.

Torgerson J. R., Branning D., Monken C. H., and Mandel L., "Experimental demonstration of the violation of local realism without Bell inequalities", *Physics Letters A* 204 (1995), 323.

Wiseman H. M., "From Einstein's theorem to Bell's theorem: a history of quantum non-locality", *Contemporary Physics* 47 (2006), 79.

6

Quantum Information, Quantum Cryptography, and Quantum Teleportation

*Any sufficiently advanced technology
is indistinguishable from magic.*
ARTHUR C. CLARKE

Quantum mechanics is magic.
DANNY GREENBERGER

6.1 Quantum Information Science

The unraveling of the quantum nature of light and matter that took place throughout the twentieth century is not only an intellectual achievement of the highest order; it led to technological innovations that affect us in everyday life. For example, the laser, which can be found everywhere from the operating room to the CD player to the transmission of information over fiber-optic networks, is the direct result of the quantum nature of atoms and of the electromagnetic field. Then there exists the field of electronics. In the "old days", prior to the 1960s, electronic devices such as radios and television sets used vacuum tubes, sometimes called valves, as the basic devices to control the flow of electrical currents. The earliest fully electronic computers, built in the late 1940s and early 1950s, were gigantic, taking up entire warehouse-sized spaces. They were not very powerful and not very fast by today's standards, nor were they very efficient, as vacuum tubes generate a lot of heat and have a tendency to burn out frequently. On the other hand, a laptop computer is much more powerful and is faster than any computer that would have been imagined possible in those days. What is behind this miniaturization and increase in speed?

Vacuum-tube devices do not rely on the quantum. They are built with nineteenth-century technology. But with the advent of quantum mechanics in the twentieth century came a fundamental understanding of not only the quantum nature of light and atoms, but also of matter in the solid state. The understanding of the quantum nature of certain kinds of crystals—silicon and germanium, for example—led to the development of new kinds of electrical materials: *semiconductors*. Semiconductors allowed for the development of new kinds of electronic devices known as transistors, which take over the operations formerly performed by vacuum tubes.

Circuits built with transistors have the advantages of being much smaller, and of generating much less heat than the corresponding vacuum tube circuits. Invented in the mid-1950s, the transistor caused a technological revolution that some refer to as the "first quantum revolution".* This revolution continued with the development of integrated circuits where entire electronic circuits can be imprinted onto a semiconducting wafer (a microchip), by a process known as known as photolithography. According to Moore's "law" (after G. E. Moore, one of the founders of INTEL), the number of transistors per microchip doubles about every 18 months, based on historical data. The "law" has held for the past 30 years, and it means that the elements of the circuits are becoming smaller and smaller and thus closer and closer together. The closer together the circuit elements, the greater the speed of devices, such as the central processing unit (CPU) of a digital computer. The first quantum revolution makes this all possible because quantum theory tells us how to build and miniaturize transistors and integrated circuits by exploiting certain quantum effects having to do with the structure of matter. However, there are no quantum coherence effects—that is, effects directly related to quantum superposition and entanglement—involved in the operation of digital devices made with integrated circuits. In that sense, their operation is just as classical as any of the earlier devices made with vacuum tubes. But as circuits become smaller and smaller, true quantum effects will start to become important. In one sense, this is bad. Inherent quantum uncertainties

* By this we mean the technological developments that occurred based on the understanding of matter brought about by quantum mechanics—the transition being a prime example. The laser is another example.

may not allow circuits to function reliably. For example, with smaller circuits, wires will be very close together, and there is the possibility of electrons tunneling from one wire to another—an unwanted effect that could foul the operation of the circuit. But in another sense, quantum uncertainty might turn out to be a tremendous advantage. In fact, the "second quantum revolution"—underway for only the past two decades or so—consists of the development of information-processing technology that directly takes advantage of quantum superposition and entanglement. This new field, broadly known as quantum information science (QIS), encompasses several overlapping subdisciplines: quantum computing, quantum key distribution (also known as quantum cryptography), and quantum metrology (the measurement of small parameters and weak forces with ultra-high precision).

Information is most conveniently represented by binary "bits", usually denoted with 0 and 1. Numbers and other kinds of information can be represented by strings of 0s and 1s, and a (classical) computer can manipulate these strings of bits, or store them in a register. The 0 and 1 are characterized by different voltage levels in a classical computer. In QIS, bits of information are encoded in quantum states, which themselves can be in superposition states and/or entanglement states. Generically, the 0 and 1 quantum bits are represented as the *qubits* (quantum bits) $|0\rangle$ and $|1\rangle$, respectively. Of course, in an actual quantum-processing device, the qubits must be represented by the states of the quantum system being used. For example, the photon polarization state $|H\rangle$ could represent the $|0\rangle$ qubit, while $|V\rangle$ could represent the $|1\rangle$ qubit. But qubits, being quantum bits, can be put into superposition states such as $|+\rangle = (|0\rangle + |1\rangle)/\sqrt{2}$ and $|-\rangle = (|0\rangle - |1\rangle)/\sqrt{2}$. For the case where $|0\rangle$ and and $|1\rangle$ represent vertically and horizontally photons, respectively, the states $|+\rangle$ and $|-\rangle$ represent photons polarized at $\pm 45°$, respectively.

The importance of QIS is that new ways of manipulating information, through quantum gates, allow for the processing of information at a high speed not possible with any kind of classical processor. To see why this speed-up could be of importance, consider the field of cryptography: the science of making and breaking codes. Most cryptographic systems in use today base their security (keeping the key secret) on the difficulty of finding the prime factors of very large integers. The prime numbers are those integers having only 1 and themselves as divisors. The numbers 1, 2, 3, 5, 7, 11, and so on, are prime numbers. The set of prime factors of the very large integer is the set of all prime numbers that are divisors of the integer. If they are known, the key to the encryption is known—bad news should an eavesdropper, Eve, be good at factoring. The very large integer is publicly available and would be known to Eve, but because finding its prime factors is an extremely difficult computational problem, the key is secure for all practical purposes. In fact, as the numbers used in practical cryptographic systems are very large, the time it would take even a very fast contemporary digital supercomputer to find the prime factors using known algorithms would be greater than the age of the universe—which is about 14 billion years! The time required to break a code by this method is said to be "exponential" in the length of the number.

Of course, it is always possible that a new algorithm could be found that could speed up the calculation manyfold on a classical computer. But in 1994 Peter Shor, then of IBM, showed that a quantum computer, if such a thing could be built, could obtain prime factors over a much shorter timescale than is possible with known algorithms running on ordinary digital computers. Shor's factorization algorithm cannot be implemented on a classical computer. A quantum computer, on the other hand, because of quantum superpositions, would operate in a manner entirely different from that of a classical computer. In the latter, the operations required to perform a calculation are essentially sequential, but in the former, quantum superposition allows for highly parallel processing in a way not possible in any kind of classical computing machinery. In a classical computer at any time during processing, the state of a register will be some definite configuration of bits of information, whereas in a quantum computer a register is generally in a superposition of such definite configurations, meaning that the configuration of the register is objectively indefinite.

Unfortunately, or fortunately, depending on how one looks at it, a large-scale quantum computer has yet to be built, nor is one on the horizon for the foreseeable future. Shor's factorization algorithm has been implemented on a small-scale quantum computer based on a process called *nuclear magnetic resonance*—the same process behind magnetic resonance imaging (MRI)—to find the prime factors of 15 (3 and 5).

However, this process cannot be scaled up to factor the very large numbers used for encryption. More recently, two groups—Jian-Wei Pan's group at the Hefei National Laboratory, and Andrew G. White's group at the University of Queensland—have performed optical versions of Shor's algorithm and have made claims that the techniques they used can be scaled up. However, in general, large-scale quantum computing will be difficult no matter how it is implemented because the particles and interactions used to process quantum information will be subjected to random and uncontrollable influences of the environment surrounding the device. These environmental effects tend to degrade, or decohere, quantum superpositions, thus spoiling the very essence of what makes quantum computation theoretically possible in the first place. This degradation of quantum superpositions—a process known as *decoherence*—turns quantum states into classical states. We shall have more to say about decoherence in the next chapter.

6.2 Quantum Key Distribution

Quantum key distribution (QKD), or quantum cryptography, is a completely secure method of secret communication, and will remain so even if a quantum computer becomes available. The basic idea is that Alice and Bob can establish a cryptographic key using quantum states—usually, in practice, polarization states of single photons—which is unbreakable by an eavesdropper (Eve). Because quantum superposition is involved, any attempt by Eve to intercept and measure the quantum superposition states and obtain the key will inevitably alter the state itself, which she must still send on to Bob. So, because of a prearranged protocol between Alice and Bob for establishing the key, Bob will know, from certain deviations of his measured results from what he expects, that someone is trying to eavesdrop. Because Eve also does a measurement and collapses a superposition onto a definite state, she does not know, and cannot know, the state of the photon sent by Alice, and thus cannot know what state to forward to Bob. Quantum key distribution maintains security because of the effects of measurement on quantum superposition states.

To see how this works, we first must say a little more about bits and binary numbers. The digits 0 and 1 form the elements of a number system known as base 2, or as the binary system. In everyday life we

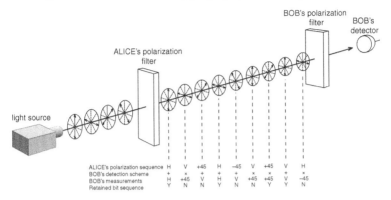

ALICE's polarization sequence	H	V	+45	H	−45	V	+45	V	H
BOB's detection scheme	+	×	+	+	+	×	×	+	×
BOB's measurements	H	+45	V	H	V	+45	+45	V	−45
Retained bit sequence	Y	N	N	Y	N	N	Y	Y	N

Figure 6.1 Sketch of the BB84 protocol for QKD using polarized photons. Alice and Bob have polarizers that can be rotated so as to produce H, V, $+45°$, and $-45°$ polarized photons. Alice sends Bob a sequence of photons of random polarization. Bob randomly performs measurements of the " $+$ " or " $\times$ " type, as described in the text.

normally work in base 10, but digital computers store and process information in base 2. In base 10 we count according to 0, 1, 2, 3, 4, ..., while in base 2 we count according to 0, 1, 10, 11, 100, ... Thus we have the equivalences $0_{10} = 0_2$, $1_{10} = 1_2$, $2_{10} = 10_2$, $3_{10} = 11_2$, $4_{10} = 100_2$, and so forth. We can use strings of bits to represent all kinds of information.

Now suppose Alice has a number that must be kept secret but which she wants to communicate to Bob. To make it simple, suppose the number is 5 in base 10, which turns out to be 101 in base 2. Alice needs to encrypt the number in some way and send it to Bob, who then performs the reverse operation, the decryption, to obtain the secret number 5. Alice then encrypts the number by adding to it, using binary arithmetic, a three-bit string which is known as the *key*. Alice and Bob have somehow agreed upon the key, which they must keep secret. So Alice, using binary arithmetic, wherein $1 + 1 = 10$ (in base 2), adds the key to the binary representation of 5 to obtain $101 + 001 = 110$. She then sends, possibly over a public channel, the number 110 to Bob who, knowing the key, subtracts it from 110 to obtain 101. The trick, of course, is for Alice and Bob to establish the key and to do it in such a way that an eavesdropper, Eve, cannot access it, or better, reveals her attempts to access it. And that brings us to quantum key distribution.

We shall discuss a QKD distribution scheme (or protocol) known as BB84, named after Charles H. Bennett (of IBM) and Gilles Brassard (of the University of Montreal), who presented the protocol in 1984. This was the first QKD protocol ever devised. We shall use polarization states of the type used in the previous chapter. The polarizations H and V are chosen to represent bits with the binary values 0 and 1 respectively, and we can similarly assign the $\pm 45°$ polarizations to represent the 0 and 1 respectively. And we must not forget that these photon polarization states are superpositions of each other, as given by eqns. (5.18)–(5.21). The set-up for this protocol is illustrated in Figure 6.1.

We assume that Alice has a source of photons whose polarizations she can choose at random. She sends to Bob a sequence of photons that are randomly polarized as either H, V, $+45°$, or $-45°$. Bob can perform one of two types of measurement: one which discriminates between H and V polarizations, which we denote as " $+$ " measurement, and the other which discriminates between the $\pm 45°$ polarizations, which we denote as a " $\times$ " measurement. Bob keeps a record of the results of measurements. Over a non-secure public channel—a telephone line perhaps—Bob tells Alice what kind of measurement he has performed on each photon, but does not reveal the results. Alice then tells Bob which of his measurements were of the correct type. For example, if Alice sends Bob a photon with polarization H, and she learns that he performed a " $+$ " measurement, she will tell him that he made a correct measurement. By making a " $+$ " measurement he will learn that the polarization of the photon is H, and thus Alice and Bob both know the polarization of the photon without having revealed it. On the other hand, if Bob performs a " $\times$ " measurement instead, then, because the polarization state $|H\rangle$ is a superposition of the $|\pm 45°\rangle$ states, he will randomly get either $\pm 45°$. He will not be able to distinguish between the states $|H\rangle$ and $|V\rangle$, because the latter is superposition of the states $|\pm 45°\rangle$. So in this sense, Bob has made the wrong measurement, and the result must be discarded. Similarly, if Alice sends a $+45°$ polarized photon and Bob makes a " $\times$ " measurement, the results are kept, while if he makes a " $+$ " measurement the results are discarded. Then, with the results that are kept, which are known to both Alice and Bob, those that are H polarized are assigned to value 0, those that are V are assigned 1, $-45°$ is assigned 0, and $+45°$ is assigned 1.

As an example, suppose Alice sends Bob a sequence of photons with the polarizations

H   V   +45° H   −45° V   +45° V   H.....

and Bob makes the sequence of measurements

+   ×   +   +   +   ×   ×   +   ×.....

and records his results

H   −45° V   H   V   +45° +45° V   −45°...

He then tells Alice the type of measurement he made in each case, and she tells him which of those are the correct type. These are

Yes   No   No   Yes   No   No   Yes   Yes   No...

Finally, from those that are the correct type, the polarizations are used to assign the numbers

0   −   −   0   −   −   1   1   −

Thus, distilled out of this process is the string of bits 0011..., which constitutes the key. Note that the first three bits are 001—precisely the key we discussed to encrypt the number 5 in binary—and we have just discussed precisely how Alice and Bob can arrive at this key. For a different sequence of photon polarizations sent to Bob upon which he performs one or the other type of measurement, the string of bits will be different and, thus, so will the key.

The security of this QKD protocol can be understood as follows: Suppose Eve intercepts a photon from Alice. She will have no idea of its polarization and can perform only a "+" or "×" measurement. Suppose Alice sends an *H* photon to Bob, but Eve intercepts it, does a "×" measurement, and obtains the result +45°. In order not to reveal her presence overtly, she must send a photon to Bob. But what polarization will she assign to it? Suppose she sends a +45° photon and Bob performs a "+" measurement. He could obtain *H*, which would be correct, but he could also obtain *V*, which would not be correct. He would tell Alice that he made a "+" measurement, and she would tell him that it was the correct type. Because she sent an *H* photon, Alice would assign it a 0. But if Bob obtained *V* he would assign it a 1. The net result of Eve's activity is to randomize the strings of bits. Thus, if Alice and Bob publicly compare random subsets of their data, errors will be discovered which reveal the presence of Eve.

They then must discard all their data and start all over. In reality, even when there is no eavesdropper, there will be errors in the data, and only when the errors exceed a certain level—say 10%—does one have to seriously suspect the presence of an eavesdropper.

We have presented here only the bare bones of a QKD protocol. Implementation is much more complicated than we have revealed, there being issues of detector efficiency and so forth. Furthermore, the BB84 protocol is only one, and perhaps the least sophisticated, of a number QKD protocols, most of which rely on entanglement for their security. They may involve states of the Bell type discussed in Chapter 5. In order to secretly communicate over large distances using entangled photons, entanglement between two photons, one photon for Alice and the other for Bob, must be maintained over these distances. As of the time of writing, the record for efficient entanglement distribution is 200 km, achieved by a group at Toshiba Laboratories in Cambridge, UK, and at the NTT Corporation in Kanagawa, Japan. But a group from the University of Geneva and Corning, Incorporated, have performed high-rate QKD with very low-loss optical fibers, made by Corning, over a distance of 250 km. The protocol used did not involve entanglement.

As we have mentioned, QKD has already gone commercial. Furthermore, there have been some interesting public demonstrations— one being in connection with a recent Swiss election, and another to make a transaction in a Swiss bank. Quantum key distribution appears to be coming of age.

6.3 Quantum Teleportation

I'm doing the best I can, Captain Kirk!

One of the more exotic features to have emerged from the field of quantum information science is the ability to transfer an unknown quantum state from one location to another, possibly very distant, location by using quantum entanglement. This procedure is known as *quantum teleportation*, and was proposed by C. H. Bennett and collaborators. The particle carrying the original state is itself *not* teleported; in fact, it is generally destroyed during the process. Rather, it is the state of the particle that becomes teleported. We shall here discuss this phenomenon in detail, assuming that the particles carrying all the relevant-state information are polarized photons. But first we need to make just a few remarks about a special kind of projective measurement that is needed to carry the teleportation protocol.

We start by assuming that Alice wants to teleport to Bob an unknown one-photon polarization state $|\psi\rangle_X = c_H|H\rangle_X + c_V|V\rangle_X$, the numbers c_H and c_V being unknown to Alice. The subscripted X is associated with the photon whose state is unknown (because c_H and c_V are unknown). A source of entangled light produces a two-photon polarization entangled state that is shared by Alice and Bob, as indicated in Figure 6.2. We take this state to be

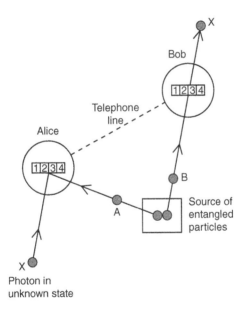

Figure 6.2 Sketch of the central idea behind quantum teleportation. Alice possesses a photon in an unknown state and one photon of an entangled two-photon state. Bob possesses the other photon of the entangled state. Alice performs a joint measurement on her photons that results in the detection of one of the four states $|\Phi_1\rangle, |\Phi_2\rangle, |\Phi_3\rangle$, or $|\Phi_4\rangle$. She then communicates her result over a classical channel (a telephone line) to Bob. Bob then knows what to do to place his photon into the state being teleported.

$$|\Psi\rangle_{AB} = \frac{1}{\sqrt{2}}[|H\rangle_A|V\rangle_B - |V\rangle_A|H\rangle_B], \tag{6.1}$$

though it could be any two-photon entangled state. This is a super-position of states wherein one photon is polarized horizontally and the other vertically. We can juxtapose the state to be teleported, and the shared entangled state as $|\Phi\rangle = |\psi\rangle_X|\Psi\rangle_{AB}$ which can, with just a little algebra, be written out as

$$\begin{aligned}|\Phi\rangle_{XAB} = \frac{1}{\sqrt{2}}[&c_H|H\rangle_X|H\rangle_A|V\rangle_B - c_H|H\rangle_X|V\rangle_A|H\rangle_B \\ &+ c_V|V\rangle_X|H\rangle_A|V\rangle_B - c_V|V\rangle_X|V\rangle_A|H\rangle_B].\end{aligned} \tag{6.2}$$

Remember that Alice has in possession the unknown state (X) to be teleported, and one part of the entangled two-photon state (A), while Bob has only the other part (B) of the entangled state.

We now introduce a set of four states involving the photon in the unknown state and the photon in state A. These are:

$$|\Phi_1\rangle \equiv \frac{1}{\sqrt{2}}[|V\rangle_X|H\rangle_A - |H\rangle_X|V\rangle_A], \tag{6.3}$$

$$|\Phi_2\rangle \equiv \frac{1}{\sqrt{2}}[|V\rangle_X|H\rangle_A + |H\rangle_X|V\rangle_A], \tag{6.4}$$

$$|\Phi_3\rangle \equiv \frac{1}{\sqrt{2}}[|H\rangle_X|H\rangle_A - |V\rangle_X|V\rangle_A], \tag{6.5}$$

and

$$|\Phi_4\rangle \equiv \frac{1}{\sqrt{2}}[|H\rangle_X|H\rangle_A + |V\rangle_X|V\rangle_A], \tag{6.6}$$

where the symbol $\equiv$ means "defined as". These are often called *Bell states* because of their close connection to Bell's theorem (all of them can serve as states that bring about violations of Bell's inequality). From eqns. (6.3) and (6.4) we have that $|H\rangle_X|V\rangle_A = (|\Phi_1\rangle + |\Phi_2\rangle)/\sqrt{2}$ and $|H\rangle_X|V\rangle_A = (|\Phi_2\rangle - |\Phi_1\rangle)/\sqrt{2}$, while from eqns. (6.5) and (6.6) we have that $|H\rangle_X|H\rangle_A = (|\Phi_3\rangle + |\Phi_4\rangle)/\sqrt{2}$ and $|V\rangle_X|V\rangle_A = (|\Phi_3\rangle - |\Phi_4\rangle)/\sqrt{2}$. Using these results in eqn. (6.2), we can rewrite the state $|\Phi\rangle_{XAB}$ in terms of the Bell states as

$$|\varPhi\rangle_{\text{XAB}} = \frac{1}{2}[|\varPhi_1\rangle(c_\text{H}|H\rangle_\text{B} + c_\text{V}|V\rangle_\text{B}) + |\varPhi_2\rangle(-c_\text{H}|H\rangle_\text{B} + c_\text{V}|V\rangle_\text{B})$$
$$+ |\varPhi_3\rangle(c_\text{H}|V\rangle_\text{B} + c_\text{V}|H\rangle_\text{B}) + |\varPhi_4\rangle(c_\text{H}|V\rangle_\text{B} - c_\text{V}|H\rangle_\text{B})].$$

$$(6.7)$$

We now assume that Alice makes measurements on the A photon and on the photon in the unknown state $|\psi\rangle_\text{X}$ that distinguish between each of the four states $|\varPhi_1\rangle, \ldots, |\varPhi_4\rangle$. Suppose she detects $|\varPhi_1\rangle$. Then Bob's photon is projected into the state $c_\text{H}|H\rangle_\text{B} + c_\text{V}|V\rangle_\text{B}$. So, if Alice rings up Bob and tells him that she has detected the state $|\varPhi_1\rangle$, he then knows that his photon is already in the state of the original photon, that state being $c_\text{H}|H\rangle_\text{X} + c_\text{V}|V\rangle_\text{X}$. The photon involved is not the same one that was in the original unknown state, but the numbers c_H and c_V are the same as for that state and are still unknown. Notice that the original state being teleported—and indeed, the photon in that state—are generally destroyed by the measurement performed by Alice. A distant joint measurement by Alice on the original photon in the state to be teleported, and on the photon in a shared entangled state yielding a particular result, projects Bob's photon into the same state as the original photon. Quantum teleportation is another non-local "spooky-action-at-a-distance" effect of the type that bothered Einstein.

As can be seen from eqn. (6.7), when Alice performs a joint measurement on the X and A photons she has a one-in-four chance of getting any of the states $|\varPhi_1\rangle, \ldots, |\varPhi_4\rangle$. Suppose she gets $|\varPhi_2\rangle$ and communicates this result to Bob. He then knows that his photon has been projected into the state $-c_\text{H}|H\rangle_\text{B} + c_\text{V}|V\rangle_\text{B}$. But this is not the original state, due to the presence of the minus sign. To regain that state, Bob must perform a transformation on his photon such that $|H\rangle_\text{B} \rightarrow -|H\rangle_\text{B}$ and $|V\rangle_\text{B} \rightarrow |V\rangle_\text{B}$. Similarly, if Alice detects $|\varPhi_3\rangle$, Bob's photon will be projected into the state $c_\text{H}|V\rangle_\text{B} + c_\text{V}|H\rangle_\text{B}$ and Bob will know that he has to make the transformations $|V\rangle_\text{B} \rightarrow |H\rangle_\text{B}$ and $|H\rangle_\text{B} \rightarrow |V\rangle_\text{B}$. Finally, if Alice detects $|\varPhi_4\rangle$, Bob's photon will be projected into the state $c_\text{H}|V\rangle_\text{B} - c_\text{V}|H\rangle_\text{B}$ and Bob will know that he has to make the transformations $|V\rangle_\text{B} \rightarrow |H\rangle_\text{B}$ and $|H\rangle_\text{B} \rightarrow -|V\rangle_\text{B}$. Thus, in all cases, the original state being teleported can be reconstructed by Bob. The manipulations required of him on the photon polarization states can be performed by using various readily available optical devices that adjust phases and alter polarization.

The phenomenon we have just described is called quantum teleportation because, in principle, Alice and Bob can be widely separated—just what you might expect if you are familiar with the *Star Trek* series, though they actually refer to their device as a "transporter". Ironically, it turns out that teleportation on that show was the result of production cost-cutting measures—an "invention" implemented in order to avoid the expense of repeated filmings of shuttle-craft landings on alien worlds. Teleportation, as portrayed on *Star Trek*, is similar to quantum teleportation, but also a little different. The *Star Trek* writers seem to have given their transporter the ability to send the material of an object which then gets reconstructed: the object or person to be transported is dematerialized, the material is transported, and the object reconstructed at a remote location. It would really not be necessary to transport the actual material: one could simply make a copy of the original, assuming one knew exactly its composition and structure. It makes no difference that the copy might be constructed of different atoms than the original, as all atoms of a specific kind are identical: one carbon atom is indistinguishable from any other. In contrast to teleportation as portrayed on Star Trek, quantum teleportation does not transport the atoms or photons maintaining the unknown quantum state. In fact, the original unknown polarization state is destroyed by the teleportation process, and the state reconstructed by Bob is identical to the original state, even though Bob still does not know what that state is.

6.4 The Experiment

We now describe the teleportation experiment carried out by Zeilinger's group in 1997. A sketch of the experimental design is given in Figure 6.3. The crystal is used to produce both the state to be teleported and the entangled state to be shared by Alice and Bob, who are represented by the two detection set-ups in the figure. A pulse of ultraviolet (UV) laser light enters the crystal from the left, and can produce, by the process shown in Figure 6.3, a pair of polarization entangled photons in beams A and B; that is, the state $|\Psi\rangle_{AB}$ given above. Beam A is reflected to Alice, while B continues on to Bob. Most of the UV photons in the laser pulse pass through the crystal and encounter a mirror, which reflects them back through the

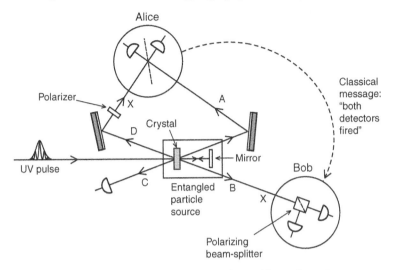

Figure 6.3 A sketch of the experiment performed by Zeilinger's group, as discussed in the text.

crystal where it produces yet another entangled pair of photons in the beams labeled C and D in the state

$$|\Psi\rangle_{CD} = \frac{1}{\sqrt{2}}[|H\rangle_C|V\rangle_D - |V\rangle_C|H\rangle_D].$$

We do not use the entanglement in this state directly. Beam C goes to a detector which has a polarization filter in front of it. The filter can be set so that only, say, the horizontally polarized photon of the C beam reaches the detector. When the detector clicks, the experimenter knows that the D beam has a vertically polarized photon; that is, we have the state reduction $|\Psi\rangle_{CD} \xrightarrow{\text{detection of } |H\rangle_C} |V\rangle_D$. Thus,

his detection of a photon in C signals, because of the entanglement, the presence of the partner photon in beam D. Furthermore, the detection of the photon in C "frees up" the photon in D; that is, the photon in D is no longer entangled. This photon will carry the state to be teleported. It will be prepared by the polarizer, set at an arbitrary angle, and placed after the mirror which reflects the photon to Alice's beam-splitter. The polarizer causes the transformation $|V\rangle_D \xrightarrow{\text{polarizer}} |\psi\rangle_X = c_H|H\rangle_X + c_V|V\rangle_X$, where the numbers c_H and c_V

depend only on the angle of the polarizer. In principle, Alice does not know the angle-setting of the polarizer, though of course, someone in the laboratory—the person who set the angle of the polarizer—does know precisely in which state the photon is prepared. Notice from Figure 6.3 that Alice has only a beam-splitter and two detectors, not the polarizer.

At this point, Alice, as you will recall, has to perform a measurement to detect one of the states $|\Phi_1\rangle, \ldots, |\Phi_4\rangle$. Because these are two-photon states, it is not easy to perform measurements to distinguish between them, and this had never been accomplished prior to this experiment. Let us see how this can be done.

Alice has a beam-splitter, and allows each of the photons from the X and A beams to fall simultaneously on opposite sides of it. Let us consider first the states $|\Phi_3\rangle$ and $|\Phi_4\rangle$. If you look back at the composition of these states you will see that they are both superpositions of the combinations $|H\rangle_X|H\rangle_A$ and $|V\rangle_X|V\rangle_A$. The photons in each of these component beams are identical, as they have the same polarizations (either HH or VV), and thus the photons will go out along the same paths if they strike the beam-splitter on opposite sides simultaneously—this being the Hong–Ou–Mandel effect discussed in Chapter 4. For that reason it will be impossible to distinguish between the states $|\Phi_3\rangle$ and $|\Phi_4\rangle$. It turns out that the photons in the state $|\Phi_2\rangle$ falling on the beam-splitter also results in the photons going in the same path toward a detector, even though the photons are not identical in this case. We can see how this happens by expressing $|\Phi_2\rangle$ in the $|\pm\rangle$ polarization states, so that

$$|\Phi_2\rangle = \frac{1}{\sqrt{2}}[|+\rangle_X|+\rangle_A - |-\rangle_X|-\rangle_A].$$

Each term in this superposition involves identical photons. Thus, when they strike the beam-splitter they will go in the same direction: the Hong–Ou–Mandel effect once again.

Hence, in the Zeilinger experiment there can be no distinction between $|\Phi_2\rangle$, $|\Phi_3\rangle$, and $|\Phi_4\rangle$. Therefore, our only hope for teleportation is to be able to detect $|\Phi_1\rangle$! Well, the "$-$" sign between its components, again $|H\rangle_X|V\rangle_A$ and $|V\rangle_X|H\rangle_A$, causes the cancellation of all the possible outcomes wherein the photons appear in the *same* beams. That is, the two photons in this case must take different paths out of the beam-splitter, and thus both detectors will fire. Therefore, the Bell state $|\Phi_1\rangle$ can be distinguished from all the other Bell states. Because all four are equally likely to be present, about 25% of all the runs of the experiment will result in the detection of $|\Phi_1\rangle$. Obviously, the other 75% of the runs, where only one detector clicks, must be discarded.

So, when Alice detects $|\Phi_1\rangle$ and communicates her result to Bob, Bob will know that he already has the unknown state: he does not need to do anything, as per the discussion in the previous section. In fact, the moment Alice detects $|\Phi_1\rangle$, Bob's photon is instantly projected into the state $c_H|H\rangle_B + c_V|V\rangle_B$, which is a *copy* of the unknown state $|\psi\rangle_X = c_H|H\rangle_X + c_V|V\rangle_X$. It remains, then, to show experimentally that these states really are copies of each other.

In the experiment, the polarizer used to generate the "unknown" state was set to $+45°$, meaning that $|\psi\rangle_X = |+\rangle_X = (|H\rangle_X + |V\rangle_X)/\sqrt{2}$. Thus, when Alice detects $|\Phi_1\rangle$, Bob has the state $|+45°\rangle_B = \frac{1}{\sqrt{2}}(|H\rangle_B + |V\rangle_B)$. To verify this, Bob has a polarizing beam-splitter which, as indicated in Figure 6.2, has two output channels—one for $+45°$ polarization and the other for $-45°$. Only the detector in the $+45°$ beam should click every time after Alice detects $|\Phi_1\rangle$, which was observed in the Zeilinger experiment. This experiment, like all experiments, was not 100% "clean"—meaning that there are always experimental uncertainties of various kinds. The Zeilinger experiment had a "fidelity" to the ideal of perfect results of about 70%, and more recent experiments have attained fidelities of about 90%.

In the time since this experiment was performed, techniques have been found for performing quantum state teleportation using any one of the four Bell states.

In the experiment just described, the state to be teleported was a photonic state. However, it is also possible to teleport quantum states of matter. Such has been done in a recent experiment by a group headed by Christopher Monroe at the University of Maryland, where they teleported a state of a trapped atomic ion to another atomic ion held 1 meter away in a different trap. The states teleported were of states of matter (an ion), but the teleportation was accomplished using photons. We shall not go into the details of this experiment here, but we duly note the experimental teleportation of the quantum states of matter—a feat that brings us even closer to the advanced technology seen on *Star Trek*.

Quantum teleportation involves the transfer of information about a quantum state—information that is not known to either the sender or the receiver. The transfer, in principle, can take place over a great distance. As far as we are aware, the greatest distance over which teleportation has been performed is 2 km, having been achieved by Nicholas Gisin's group in Geneva. Such a transfer is possible because of the inherent non-locality embedded in entangled states, and because of the effects of projective quantum measurements.

A possible application of quantum teleportation is to transfer quantum states to different parts of a quantum computer. That is, of course, if such a machine is ever built.

Bibliography

Bennett C. H., Brassard G., and Eckert A. K., "Quantum Cryptography", *Scientific American*, October 1992, p. 50.

Bennett C. H., Brassard G., Crépeau C., Jozsa R., Peres A., and Wooters W. K., "Teleporting an unknown quantum state via dual classical and Einstein–Poldosky–Rosen channels", *Physical Review Letters* 70 (1993), 1895.

Boschi D., Branca S., De Martini F., Hardy L., and Popescu S., "Experimental realization of teleporting an unknown pure quantum state via dual classical and Einstein–Podolsky–Rosen channels", *Physical Review Letters* 80 (1998), 1121.

Bouwmeester D., Pan J.-W., Mattle K., Eibl M., Weinfurter H., and Zeilinger A., "Experimental quantum teleportation", *Nature* 390 (1997), 575.

Brown J., *The Quest for the Quantum Computer*, Simon and Schuster, 2000.

Lloyd S., *Programming the Universe: A Quantum Computer Scientist's Takes on the Cosmos*, Vintage, 2007.

Marcikic I., de Riedmatten H., Tittel W., Zbinden H., and Gisin N., "Long-distance teleportation of qubits at telecommunication wave length", *Nature* 421 (2003), 509.

Milburn G. J., *Schrödinger's Machines: The Technology Reshaping Everyday Life*, W. H. Freeman, 1997.

Milburn G. J., *The Feynman Processor: Quantum Entanglement and the Computing Revolution*, Perseus Books, 1998.

Olmschenk S., Matsukevich D. M., Maunz P., Hayes D., Huan L.-M., and Monroe C., "Quantum teleportation between distant matter qubits", *Science* 323 (2009), 486.

Singh S., *The Code Book: The Science of Secrecy from Ancient Egypt to Quantum Cryptography*, Anchor Books, 2000.

Schrödinger's Cat and Leggett's SQUID: Quantum Effects on a Large Scale?

7.1 The Large, the Small, and the In-Between

We have been talking about some pretty weird stuff in this book: interference effects with electrons passing one at a time through a double-slit arrangement, interference effects with photons going around both arms of an interferometer one at a time, delayed choice between particle–wave detection schemes, quantum erasure and recovery of information, detection of the presence of an object without scattering photons from it, superluminal effects in photon tunneling, the ability of entangled particles to influence each other over large distances, quantum teleportation, and so on. We have consistently been able to explain, or at least describe, these effects with the principles of quantum mechanics. At the heart of quantum mechanics is the principle of superposition of distinctly different quantum states. That is, the states that compose the superposition possess definite but different values of some physical attribute. But when these states are superposed to form a new state, the attribute of the particle is no longer objectively definite. The attribute is indeterminate, and we can only give probabilities that a measurement of it yields a particular value. The state vector for a quantum system is generally a superposition of the state vectors associated with definite values for a particular attribute of the system, and measurement of the attribute causes a collapse of the superposition onto the state corresponding to the value of the attribute obtained by the measurement. The measured value of the attribute and the collapse of the state vector are irreducibly probabilistic.

The strangeness of the quantum world is perhaps mitigated by the fact that we are talking about "microscopic" things that are the behavior of electrons, atoms, photon, and so on, which are "microscopic" things on the scale of the atomic world. Our intuition is not a good guide on that scale; we simply cannot experience it directly. Nor do we ever experience macroscopic objects in superposition states. How would we know it if we ever did?!

Like Einstein, Erwin Schrödinger, one of the founders of modern quantum mechanics,* had trouble accepting the measurement aspect of the Copenhagen view of quantum mechanics, and even doubted that quantum mechanics correctly described correlations of the EPR type between particles with large spatial separations. In 1935, stimulated by the EPR argument, Schrödinger highlighted the quantum measurement problem by proposing a burlesque that can apparently lead to some absurd conclusions if one takes quantum mechanics and the Copenhagen doctrine seriously on a macroscopic level. The burlesque, known as the paradox of Schrödinger's cat, addresses, among other things, the issue of the collapse of the state vector; that is, how and when does it occur? It also leads us to think about the prospect that quantum superpositions may exist for macroscopic objects, or at least mesoscopic objects, which in turn leads us to ask why such superpositions are never seen in the everyday world. An even deeper question is this: Where is the boundary between the classical and quantum worlds? What determines when we can get away with a purely classical description of phenomena that does not involve superpositions of states and when we are forced to use a quantum description that does? Is it just a matter of size? Can there really be a clear-cut boundary, any boundary at all, between what is classical and what is quantum, if it is the case that quantum mechanics underlies all of nature?

7.2 A Quantum Burlesque: The Tale of Schrödinger's Cat

Schrödinger considered the following "diabolical device", as he referred to it, pictured in Figure 7.1. A live cat is placed in a steel box along with a radioactive atom. A Geiger counter is set up to detect the decay of the atom. The counter is connected to a relay which, when

* He discovered the wave mechanics approach to quantum mechanics in the years 1925–26, almost simultaneously with Werner Heisenberg's development of the matrix mechanics approach (see the historical outline in the Appendix). The approaches turned out to be equivalent.

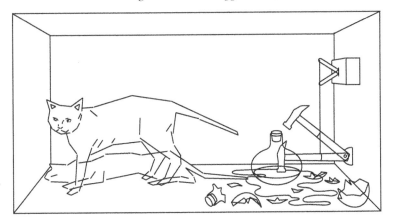

Figure 7.1 Schrödinger's cat paradox. A live cat and a radioactive atom are placed in a steel sound-proof box. The atom has a 50:50 chance of decaying within 1 hour. If the atom decays, the Geiger counter detects the decay products, and through a relay trips a switch causing a hammer to fall and break a flask containing a poisonous gas which kills the cat. After 1 hour the system will be in a superposition of the states where the atom has not decayed and the cat is alive, and the atom has decayed and the cat is dead. The state of the macroscopic object—the cat—will be objectively indefinite until someone opens the box to examine the condition of the cat. From Bryce DeWitt, "Quantum mechanics and reality", *Physics Today* (February 1970). (Used with permission from the American Institute of Physics.)

activated by the detection of the atomic decay, causes a hammer to fall and break a flask containing cyanide, a poisonous gas. Obviously, if the flask is broken, the cat will be killed (our apologies to cat-lovers). Radioactivity is a probabilistic process, and we shall assume that our radioactive atom has a 50:50 chance that it will have decayed after 1 hour. So the cat, atoms, counter, and flask are enclosed in the steel box, and we let 1 hour pass. We cannot see into the box, and, furthermore, we assume that it is sound-proof, so we cannot hear the hammer fall and break the flask.

Now, at the end of the hour, if we treat both the atom and the cat quantum-mechanically, the cat-plus-atom system should be in the state

$$|\psi\rangle = \frac{1}{\sqrt{2}}[|\text{not decayed}\rangle_{\text{atom}}\ |\text{alive}\rangle_{\text{cat}} + |\text{decayed}\rangle_{\text{atom}}\ |\text{dead}\rangle_{\text{cat}}].$$

This is a state consisting of live and dead cats entangled with a decayed or not-decayed atom—a "quite ridiculous situation", according to Schrödinger. The microscopic state of the atom is entangled with macroscopically distinguishable (dead or alive) states of the cat.[†] According to Copenhagen, after 1 hour, the atom–cat state is objectively indefinite. Furthermore, if we look into the box after 1 hour we shall surely find that either the cat is still alive, or is dead. If the former, we know that the atom has not decayed, whereas for the latter it has decayed. However, according to Copenhagen, it is the act of observing the state of the cat that collapses the state $|\psi\rangle$ onto either $|\text{not decayed}\rangle_{\text{atom}} |\text{alive}\rangle_{\text{cat}}$ or $|\text{decayed}\rangle_{\text{atom}} |\text{dead}\rangle_{\text{cat}}$, and that before the observation, the condition of the atom–cat system (cat alive, atom not decayed; cat dead, atom decayed) is objectively indeterminate. We are perhaps, by now, accustomed to the idea of microscopic particles being in states where some property is objectively indefinite. The entangled state involving a microscopic particle, an atom, and a macroscopic object, the hapless cat that can be in macroscopically distinguishable states (alive or dead), is essentially a straightforward extension of quantum phenomena of the microworld into the macroworld. If we accept quantum mechanics, we also have to accept that a macroscopic object could have objectively indefinite properties with respect to some macroscopic observable: *Is the cat alive or dead?* The idea that a large object can be in an indefinite state seems absurd on the face of it. That is the point of the burlesque. Schrödinger's cat paradox highlights the problem of locating the divide between the quantum and classical worlds required by the Copenhagen interpretation, showing that it is not well-defined.

The notion of the collapse of any state vector upon measurement is not even described by the quantum-mechanical formalism itself. For an undisturbed system, its quantum states smoothly evolve in time according to standard quantum theory. But the collapse of a state vector upon the measurement of some observable property, which involves a disturbance by an external measurement apparatus, is assumed to be abrupt—that is, sharply discontinuous—but no such

---

[†] We are not allowing for the possibility that there may be cases where the states of being alive or dead are not macroscopically distinguishable, though we can think of examples of such. Recall that Dorothy Parker, on hearing of the death of Calvin Coolidge, asked "How can they tell?"

changes are described within the theory. So, the collapse of state vectors is really an add-on. But where, or when, does the collapse occur?

To see the extent of the difficulty, let us consider how the collapse of a state vector is supposed to occur in the context of the business with Schrödinger's cat. Perhaps we need to take into account the quantum state of the Geiger counter, so that our state really should be

$$|\psi\rangle = \frac{1}{\sqrt{2}}[|\text{not decayed}\rangle_{\text{atom}} \; |\text{not triggered}\rangle_{\text{counter}} \; |\text{alive}\rangle_{\text{cat}}$$
$$+ \; |\text{decayed}\rangle_{\text{atom}} \; |\text{triggered}\rangle_{\text{counter}} \; |\text{dead}\rangle_{\text{cat}}]$$

Following this reasoning, we should next put in the quantum state of the hammer—that is, the states where the hammer has not fallen and the state where it has—to obtain

$$|\psi\rangle = \frac{1}{\sqrt{2}}[|\text{not decayed}\rangle_{\text{atom}} \; |\text{not triggered}\rangle_{\text{counter}} \; |\text{not fallen}\rangle_{\text{hammer}} \; |\text{alive}\rangle_{\text{cat}}$$
$$+ \; |\text{decayed}\rangle_{\text{atom}} \; |\text{triggered}\rangle_{\text{counter}} \; |\text{fallen}\rangle_{\text{hammer}} \; |\text{dead}\rangle_{\text{cat}}].$$

We could continue with this line of reasoning and add the flask states, such as broken or unbroken, and so on. Clearly, at this point things are starting to become surreal. In fact, if we continue with this line of reasoning we obtain an infinite regression—the so-called *von Neumann catastrophe of infinite regression*.

Why catastrophe? Because at no stage does a measurement come to completion according to this scenario. The state vector never collapses. Schrödinger's cat paradigm forces us to confront the issue of just how and where (or when) the state vector collapses. Does it occur when someone looks in the box to see whether the cat is dead or alive? When the flask breaks or does not? When the hammer falls or does not? When the relay is tripped or is not? When the Geiger counter detects or does not detect the radioactive decay? Again, where is the divide between the quantum and classical worlds? This is the most important issue raised by the cat paradox.

One attempt at resolving the issue, proposed by Eugene Wigner,[‡] is the idea that consciousness has a different role to play in the quantum

[‡] Eugene Wigner (1902–1995) was a Hungarian physicist who emigrated to the USA in 1937 and settled at Princeton University. He made many contributions to quantum physics, especially in the area of symmetries, for which he was awarded the Nobel Prize in 1963.

mechanics, at least in regard to the issue of measurements, than does an inanimate detector. Wigner suggested that the state vector collapses when it interacts with the first conscious mind encountered. In fact, it need not be a human mind. The cat is also a conscious being, and so it could be that the state vector is brought to collapse by the cat. Wigner, in effect, suggested that the cat be replaced by a human being who is a friend of the actual observer; that is, the experimenter who looks into the box. The friend of the observer is not killed (he could wear a gas mask), but he is able to see whether or not the radioactive atom undergoes a decay within the hour. The observer then has to decide whether his friend causes the collapse of the wave function or whether or not his friend becomes entangled with the atom, and that the collapse of the atom–friend state is brought about by the observer himself. The idea that the friend does not have the same ability to bring about the collapse as does the observer himself is solipsism in the extreme, and this is what led Wigner to suggest that consciousness is ultimately what is responsible for bringing about the collapse of quantum states.

However, according to Copenhagen, collapse is supposed to occur when a *microscopic* particle interacts with a *macroscopic* detector which acts to *irreversibly* amplify the signal into the macroscopic, classical world. The imposition of the macroscopic detector to bring about collapse is an essential ingredient of the Copenhagen interpretation of quantum mechanics. Minds, human or feline, are not involved. In contrast, Schrödinger thought quantum mechanics not even correct when applied to large-scale systems, and that doubt extended to cases of entangled *microscopic* particles with large spatial separations, as in the EPR scenario discussed in Chapter 5.

As an example of an irreversible amplification upon detection, recall the experiments discussed in the previous chapters where the detection of single photons was required. The photon detectors are photomultiplier tubes wherein a single photon ejects a single electron from a metal surface (the photoelectric effect), which is then accelerated by an electric field and collides with other metal surfaces. This in turn creates more electrons in a cascading process so that the photon is indirectly detected as a pulse of electric current on a macroscopic scale. Thus, the detection of a single photon involves an amplification of the original signal. The same is true for the detection of the decay of a single radioactive atom in the cat paradigm: the Geiger counter amplifies the initial signal if present, and thus, according to Copenhagen, brings about collapse of the state vector. The key point here is that the amplification by the detector puts information into the classical world, and does so irreversibly; the information cannot be put back into the quantum world. Essentially, then, state vectors collapse upon detection by what is essentially a classical apparatus. This constitutes the Copenhagen answer to the cat paradox and, by extension, to quantum-mechanical measurements on all scales. If collapse does not occur when the initial signal of the atomic decay begins to be amplified by the various classical measurement devices, there simply will not be any conclusion to any attempt at performing a measurement. Remember that superposition in quantum mechanics represents ambiguity—a lack of definitive information. In the case of Schrödinger's cat, once information about the atomic decay enters the classical world through a classical detection device (which includes an amplifier), the ambiguity is resolved, and that is the end of the story.

It happens that a phenomenon very much like a Schrödinger-cat state has long been known, and in fact, the related experiment was performed in 1922, before the discovery of modern quantum mechanics. We refer to the Stern–Gerlach experiment. This experiment involves a beam of atoms of the element silver, and the spin of just one of the electrons attached to the silver atom. The spin of an electron is another quantum-mechanical notion. Actually, the electron acts as though its charge is circulating about its axis of spin, generating a small magnetic field similar in shape to a field produced by a simple bar magnet. We said that it "acts" this way because it is hard to imagine an object of no size rotating. Nevertheless, the spin of the electron, whatever that might mean, is intimately connected to the magnetic field that it produces. Furthermore, when the electron is in a large uniform external magnetic field, it has only two possible spin orientations: one being along the field, which we call the spin-up state, represented by $| \uparrow \rangle$, and the other being against the field, which we call the spin-down state, represented by $| \downarrow \rangle$. In a silver atom there are 47 electrons. Of those, there are 23 pairs of electrons where the spins in a pair are opposite, and thus cancel out each other. The one remaining electron carries the spin for the whole atom.

Now, if the silver atoms are in a uniform magnetic field, the atoms will experience no force. However, in a non-uniform field the atom experiences a force that depends upon the direction of the spin of the single unpaired electron. Indeed, Otto Stern and Walter Gerlach found that a beam of silver atoms passing through a non-uniform magnetic field splits into two distinct beams, and *only* two beams, as pictured in Figure 7.2. The fact that there are two beams instead of a broad continuous spread of atoms is now understood as the result of the quantum nature of the electron spin. Thus, by passing the atomic beam through the Stern–Gerlach apparatus, a fairly large-scale separation of the atomic beams (a few centimeters) can be attained. The individual atoms in the beam can be characterized by i) the spin of the electron, $|\uparrow\rangle$ or $|\downarrow\rangle$, and ii) by the direction of motion of the whole atom, which we denote by $|\text{upper beam}\rangle$ or $|\text{lower beam}\rangle$. Thus, the state of one atom after passing through the Stern–Gerlach apparatus is

$$|\psi\rangle = \frac{1}{\sqrt{2}}[|\uparrow\rangle|\text{upper beam}\rangle + |\downarrow\rangle|\text{lower beam}\rangle],$$

which is exactly of the *form* of Schrödinger's cat state. Admittedly, the atom itself is not macroscopic, though the two beams can have macroscopic separations; that is, they are macroscopically distinguishable, and are tied to the internal spin state of one of the electrons in the atom. In this sense, Schrödinger's cat is realized. The screen upon which the atoms collect serves as the detector for this experiment. Obviously, the silver atom does not have the complexity of a cat. Perhaps the state generated in the Stern–Gerlach experiment can be thought of as a Schrödinger kitten.

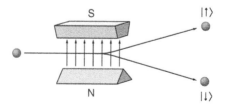

Figure 7.2 A sketch of the Stern–Gerlach experiment. A beam of silver atoms passes through the poles of a magnet shaped to produce a non-uniform magnetic field. The atoms are deflected according to the spin, up $|\uparrow\rangle$ or down $|\downarrow\rangle$, of a single electron as projected along the magnetic field.

7.3 Interference between Live and Dead Cats: Leggett's SQUID

The Schrödinger cat paradox *is* a paradox (or at least appears to be so) because the states of macroscopic objects are of considerable complexity; that is, the macroscopic cat is entangled with the states of a microscopic object: the radioactive atom. Entanglement is the essential point, and in fact, Schrödinger was the first to use the word "entanglement" in this physical context. He considered entanglement to be the distinguishing feature of quantum mechanics. Entanglement is essential to the EPR argument, Bell's theorem, the GHZ state, teleportation, and so on, as we have already seen.

Because Schrödinger's cat state involves entanglement, the states $|\text{alive}\rangle_{\text{cat}}$ and $|\text{dead}\rangle_{\text{cat}}$ do not interfere with each other. But in recent years there has been much discussion about the prospects of producing superpositions of the form

$$|a\rangle = \frac{1}{\sqrt{2}}[|\text{alive}\rangle_{\text{cat}} + |\text{dead}\rangle_{\text{cat}}],$$

so that direct interference is possible. Let us right away state that, as a practical matter, absolutely no-one imagines that objects of the complexity of a cat can ever be put into superposition states of this form—at least not for any appreciable length of time. But perhaps it is possible to find some stand-in systems of lower complexity, but with macroscopically, or at least mesoscopically, distinct quantum states playing the roles of the live and dead cat. Superpositions of live and dead cats have also come to be known as Schrödinger cat states even though they are somewhat different from those of the original discussion. Nevertheless, they would still be "quite ridiculous", as they imply that the state of a macroscopic object is objectively indeterminate, contrary to all experience.

As is always the case when confronted with superposition states, whether we are talking about suppositions of states of a microscopic particle, such as the electron passing through the double-slit apparatus as discussed in Chapter 2, or a superposition of a live and dead cat, it is tempting to interpret such a thing as simply meaning that the factual state of the system is one state or the other in the superposition, and the experiment merely reveals which. Certainly, in the case of Schrödinger cat state $|a\rangle$, this interpretation would not be terribly unsettling. But the whole point of quantum superposition is to explain interference effects, which in turn implies objective indefiniteness with respect to the relevant states of the system being superposed (here the state for the cat alive and cat dead). Therefore, if such a state as $|a\rangle$ can somehow be generated, interference effects between the live and dead cat states should be possible. The idea that states of live and dead cats can interfere with each other—that a cat can neither be objectively alive or objectively dead—surely *is* unsettling.

In 1984, Anthony Leggett (who would win the Nobel Prize in 2003 for his theoretical work in condensed-matter physics), discussed the prospects of generating cat-like states in superconducting systems involving what are called *Josephson junctions*. Superconductivity is a low-temperature phenomenon that occurs in certain materials, such as mercury, when the temperature is brought below some critical value. Electric currents—supercurrents—can flow without resistance—a feat not possible in ordinary conductors such as copper. The explanation of superconductivity is ultimately quantum-mechanical, and we shall not attempt a description here. It should be said that superconductivity ordinarily does not exhibit "quantumness"— that is, quantum coherence effects—of the type which we have been concerned with in this book, though the experiment we are about to discuss is "extraordinary" in that such an effect does appear. The current in a superconductor is macroscopically large (which is one reason why superconducting magnets are used to generate the strong magnetic field in most MRI machines), but there is nothing about the current itself that displays any of the quantum-mechanical features we have been discussing. However, there is a device known as a *superconducting quantum interference device*, or SQUID, that when operated under certain conditions can bring about these quantum effects on a fairly large scale. A SQUID is a superconducting ring containing a Josephson junction, as illustrated in Figure 7.3.

A Josephson junction is a thin slice of insulating material that interrupts the ring and provides an energy barrier to the supercurrents. The junction does not stop currents flowing around the ring; rather, the currents quantum-mechanically tunnel through the energy barrier of the insulator. With a current flowing in one direction around the ring—say clockwise—a magnetic field pointing into the page is generated in the center of the ring as shown in Figure 7.4(a).

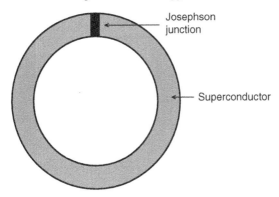

Figure 7.3 A sketch of a superconducting ring with a Josephson junction. The junction consists of a thin layer of insulation.

(All electric currents generate magnetic fields.) If the current is counter-clockwise, as in Figure 7.4(b), the magnetic field is out of the page. Sensitive magnetometers can measure the magnetic field size and direction—the direction of the field informing us of the direction of the supercurrent. A supercurrent consists of a flow of paired electrons, and is a macroscopic entity involving a very large number of electron pairs. If the ring were continuous, the magnetic field would be trapped and could never change; the magnetic field cannot penetrate the superconductor—the *Meissner effect*. But because of the thin layer of insulation that forms the Josephsen junction, the magnetic field can quantum-mechanically "tunnel" to make a quick reversal in direction. If one could somehow induce superposition states of supercurrents flowing in opposite directions in the ring, then there would be a superposition of magnetic fields of opposite directions in the center of the ring. This does *not* mean that the field is zero in the center of the ring! Because the state of the superconductor currents is

$$|a\rangle_S = \frac{1}{\sqrt{2}}[|\text{clockwise current}\rangle_S + |\text{counter-clockwise current}\rangle_S],$$

the current direction itself is objectively indefinite, and thus so is the corresponding magnetic field through the ring. Measurements of the magnetic field of the ring never yield a result of zero field. Rather, for some measurements, the field will be detected in one direction and for others in the opposite direction; the two results corresponding to

(a)

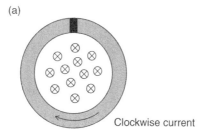

Clockwise current

(b)

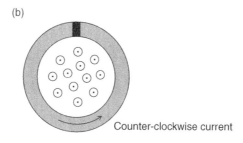

Counter-clockwise current

(c)

Indefinite
magnetic field

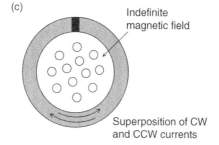

Superposition of CW
and CCW currents

Figure 7.4. Currents and magnetic fields in a superconducting ring containing a Josephson junction. (a) For a clockwise current, the magnetic field in the ring is directed into the page, as indicated by the symbol ⊗. (b) For a counter-clockwise current the magnetic field is directed out of the page, as indicated by the symbol ⊙. (c) If the current of the supercondutor is a superposition of clockwise and counter–clockwise currents, then the Magnetic field in the ring is of indefinite direction, as indicated by the symbol O.

the collapse of the state $|a\rangle_S$ onto either $|\text{clockwise}\rangle_S$ or $|\text{counter-clockwise}\rangle_S$. These possible outcomes correspond to the live or dead states of the cat. (Note that there is no question of consciousness at this level, as only inanimate devices are involved!) However, an experiment must be able to distinguish between a statistical mixture, which is essentially classical, and a superposition state, which should exhibit, in this case, complete macroscopic coherence. This can be done because for the statistical mixture the probabilities for the direction of the magnetic field are constant; that is, they do not change with time. But for a superposition, the probabilities for the magnetic field directions should change smoothly and periodically with time.

Actually, there are *two* relevant possible superposition states of the supercurrent states, one being the state $|\alpha\rangle_S$ above, where the clockwise and counter-clockwise current states are added together, and the state where they are subtracted:

$$|\beta\rangle_S = \frac{1}{\sqrt{2}}[|\text{clockwise current}\rangle_S - |\text{counter-clockwise current}\rangle_S].$$

The states $|\alpha\rangle_S$ and $|\beta\rangle_S$ are predicted to have different energies. If the current in the ring is initially, say, clockwise, then by adding $|\alpha\rangle_S$ and $|\beta\rangle_S$ we can see that the clockwise current state is just the superposition

$$|\text{clockwise current}\rangle_S = \frac{1}{\sqrt{2}}[|\alpha\rangle_S + |\beta\rangle_S].$$

As time goes on, because of the differing energies of the state $|\alpha\rangle_S$ and $|\beta\rangle_S$, there will be smooth and continuous oscillations between the currents of the ring being clockwise and being counter-clockwise, where the rate of the oscillations is related to the energy difference. The oscillations are the result of full quantum coherence between the two macroscopically distinguishable states of the current in the ring. The point is that at certain times in the oscillations, between those times when the current is definitely clockwise and definitely counter-clockwise, the current direction is objectively indefinite.

This effect was observed in the laboratory by two groups—one at the State University of New York at Stony Brook, (Friedman, et al.) and the other at the Technical University of Delft in The Netherlands (van der Wal et al.), in 2000.[§] The current in the ring consists of the flow

[§] The work of these researchers was preceded by the important work of John Clarke and collaborators at the University of California at Berkeley (see the bibliography for this chapter), who reported observing macroscopic quantum tunneling in a Josephson junction in 1988.

of about 1 billion (10^9) electrons, making them truly macroscopic. Thus, it seems that under carefully controlled conditions, the story of Schrödinger's cat is not a paradox at all; it is a *phenomenon*. In reality, there are no paradoxes in nature.

7.4 Decoherence and the Quantum/ Classical Divide, or Why are Sightings of Schrödinger's Cat so Rare?

The experiments described in the previous section demonstrate quantum interference effects on a fairly large scale. This begs the question: Why do we not see the effects of superposition states in the everyday world? This question, or the answer to it, really gets to the heart of the problem concerning the divide between the quantum and classical words. Superposition states are routinely produced in experiments with atoms, electrons, and photons, but, of course, atoms, electrons, and photons are of the microworld—the domain we normally think of as the quantum world, the world where the ordinary rules of logic break down, as we have been describing throughout this book. Virtually every experiment we have discussed up to the present chapter involves only the microworld. As we progress to larger entities, it becomes much harder to manufacture superposition states like those described in the previous section. Why?

Well, the answer has to do with the uncontrollable interactions that a quantum system experiences with the rest of the world—that is, the part of the world directly surrounding it, which we henceforth call *the environment*. No quantum system is truly isolated from the rest of the universe, though we can sometimes minimize the effects of the environment. For a small system, environmental effects can be ignored altogether over the relevant timescale of an experiment. But for larger-scale systems, such as for the interfering cat states of SQUID, the environment has the effect of destroying the quantum coherence of the superposition state at short times. Quantum coherence for the state of the supercurrents can be maintained in such experiments only for short times. The destruction of quantum coherence is known as *decoherence*. The notion of decoherence due to environmental interactions was brought into quantum mechanics by the efforts of Wojciech H. Zurek (of Los Alamos National Laboratory) and many others, beginning in the 1970s and extending through the 1980s. As the name *decoherence* implies, quantum coherence is no longer present. Decoherence occurs because when the system of interest interacts with its environment they (the system and its environment) generally become entangled. But the environment is an enormously large system which is impossible to study in detail.

To acquire a sense of what this is all about, suppose that the state of the current in the SQUID is prepared in the state $|a\rangle_S$ discussed above, and that environment is at the time in the state $|\Phi_0\rangle_E$. So the combined state of the two systems at this time is

$$|a\rangle_S |\Phi_0\rangle_E = \frac{1}{\sqrt{2}} [|\text{clockwise current}\rangle_S$$

$$+ |\text{counter-clockwise current}\rangle_S] |\Phi_0\rangle_E,$$

wherein there is no entanglement, as we just have the superconductor current state times the state of the environment. Note that the superconducting current state is a superposition of clockwise and counter-clockwise supercurrents; there exist quantum coherence between those states, the direction of flow of current being objectively indefinite. But after a short time the interactions between the systems brings about the combined state

$$\frac{1}{\sqrt{2}} [|\text{clockwise current}\rangle_S |\Phi_{CW}\rangle_E$$

$$+ |\text{counter-clockwise current}\rangle_S |\Phi_{CCW}\rangle_E],$$

where $|\Phi_{CW}\rangle_E$ and $|\Phi_{CCW}\rangle_E$ are two different states of the environment correlated to the clockwise and counter-clockwise states of the current. The state of the combined system of supercurrent plus environment is the coherent quantum superposition state (that is, an entangled state), but there is no longer any quantum coherence between each of the states of the supercurrent. Because of the correlations with the states of the environment, information on the current direction is now potentially available in the macroscopic world. The situation is similar to what happens in the case of the double-slit experiment discussed in Chapter 2, where any attempt to determine the path of the electron through the slits has the effect of

destroying the quantum coherence (that is, destroying the interference). But, as we saw elsewhere in the book—specifically in Section 4.5
in the discussion of the discussion of the "mind-boggling experiment"
of Zou, Wang, and Mandel—even the potential availability of path
information is enough to destroy quantum coherence, and this is
exactly the case here. We ignore the environment altogether, as its
size would render it practically impossible to perform measurements
on it, and we end up with a statistical mixture with respect to the
direction of the supercurrents because of the loss of quantum coherence within the superconducting ring. That is, decoherence has the
effect of rendering an initial state, given as a superposition of two
macroscopically distinct supercurrents, into a classical system void of
any quantum coherence. Symbolically,

$$|a\rangle_S = \frac{1}{\sqrt{2}}[|\text{clockwise current}\rangle_S + |\text{counter clockwise current}\rangle_S]$$

$$\xrightarrow{\text{decoherence}} \rho_{\text{Mixed}} = \begin{cases} \frac{1}{2} & \text{for clockwise current,} \\ \frac{1}{2} & \text{for counter clockwise current,} \end{cases}$$

the fraction $\frac{1}{2}$ being the probability of the respective current directions
system (note that we introduced this notation for a mixed state in
Chapter 2). The preceding is not meant to be a precise statement of
the theory of decoherence, which is quite involved. Rather, it is
meant to give the flavor of it, to model it without being overly
simplistic.

The speed of decoherence increases with the size of the system of
interest. This is so because the larger and more complex the system,
the harder it becomes to isolate it from the effects of the environment. Thus, the reason we do not have Schrödinger cat-like states in
the everyday world is that, even if they could somehow be created,
they would decohere almost instantaneously. Should anyone attempt
to produce the kind of states suggested by Schrödinger with a real cat
and a radioactive atom, the virtually instantaneous loss of coherence
through unavoidable environmental interactions would render
Schrödinger's cat either objectively dead or objectively alive.

Quantum coherence can, at least in principle, be maintained in
macroscopic systems to the degree that they can be isolated from the
effects of the environment, as in the experiments with the SQUID
described above. In this sense, the quantum-classical divide is move-

able. The key is in isolating the system of interest from the rest of the universe, and it all depends on the cleverness of the experimenter to engineer the required isolation. This becomes much harder to do as the size of the system becomes larger (has more degrees of freedom).

The logical extension of this train of thought leads to the notion that, in reality, there really is no quantum-classical border. Why should there be a border if quantum physics is the foundation for everything? When a quantum system interacts with its environment, the system and the environment become entangled. Because the environment is, in effect, "the rest of the universe", it has many, many more degrees of freedom that the system under study. If the experimenter studies only the one quantum system and ignores the environment altogether after the two have interacted, then the system of interest behaves classically: The entanglement itself prevents any quantum interference (or the quantum coherence which leads to interference) between different states of the system or the environment. So, ironically, the quantum coherence in the prepared superposition of the supercurrents flowing in opposite directions disappears because of another quantum phenomenon: entanglement! It appears that Bohr's remark to Aage Peterson, quoted in Chapter 5 ("[t]here is no quantum world...") is precisely backwards—that in reality *there is no classical world*. Fundamentally, the world is quantum-mechanical, and the macroscopic world appears to be classical because of quantum entanglements involving large numbers of particles and degrees of freedom. The decoherence of quantum superposition states into statistical mixtures is the result of increasing entanglement. That is, a mixture arises through microscopic quantum-mechanical inter-actions with the states of particles constituting the environment, and that leads to entanglement, out of which emerges the classical world. Still, on a fundamental level the world is irreducibly probabilistic and quantum-mechanical. Ironically, the classical world of everyday life arises out of quantum entanglement!

Bibliography

Clark J., Cleland A. N., Devoret M., Esteve D., and Martinis J. M., "Quantum mechanics of a macroscopic variable: The phase difference of Josephson junction", *Science* 239 (1988), 992.

Friedman J. R., Patel V., Tolpygo S. K., and Lukens J. E., "Quantum superposition of distinct macroscopic states", *Nature* 406 (2000), 43.

Giulini D., Joos E., Kiefer C., Kupsch J., Stamatescu I.-O., and Zeh H.-D., *Decoherence and the Appearance of Classical in Quantum Theory*, Springer-Verlag, 2003.

Leggett A. J., "Schrödinger's cat and her laboratory cousins," *Contemporary Physics* 25 (1984), 583.

Schrödinger E., "The present situation in quantum mechanics", published in *Naturwissenschaften* in 1935, republished in J. A. Wheeler and W. H. Zurek (eds.), *Quantum Theory and Measurement*, Princeton University Press, 1983.

van der Wal C. H., ter Haar A. C. J., Wilhelm F. K., Schouten R. N., Harmans C. J. P. M., Orlando T. P., Lloyd S., and Mooij J. E., "Quantum superpositions of macroscopic persistent current states", *Science* 290 (2000), 773.

Zeh H.-D., "On the interpretation of measurements in quantum mechanics", *Foundations of Physics* 1 (1970), 69.

Zurek W. H., "Decoherence and the transition from the quantum to the classical", *Physics Today* 44, 10 (1991), 36.

8

Quantum Philosophy

The fact that an adequate philosophical interpretation [of quantum mechanics] has been so long delayed is no doubt caused by the fact that Niels Bohr brainwashed a whole generation of physicists into thinking the job was done fifty years ago.

MURRAY GELL-MANN (NOBEL PRIZE ACCEPTANCE SPEECH, 1976)

8.1 Reducing Quantum Mechanics?

In the preceding chapters we have illustrated the strangeness of the quantum world with a selection of experiments that one would have a hard time explaining in terms that would make sense in the everyday world. The structure of quantum theory reflects the nature of the quantum world as revealed by those experiments. The picture that emerges is that the quantum world is irreducibly probabilistic, non-realistic, and non-local. The fact that the quantum world does not always agree with our common-sense notions is immaterial: we do not experience that world directly, and our common-sense notions are conditioned by the classical world of everyday life. But there are still legitimate questions that can be asked regarding its interpretation.

The history of much of modern science is a history of the application of the notion of *reductionism*: the idea that complex things can be understood by reducing them to supposedly simpler interactions among their parts. It is the idea that every level of reality can be explained by simpler laws operating on a lower level of reality—that the large can be explained in terms of the small, spiraling downwards to ever decreasing scales of structure and processes. The properties of materials are determined by the properties of the molecules of which they are composed, while the properties of the molecules are determined by the properties of their constituent atoms and the nature of the interaction that hold them together; that is, the chemical bond which itself is a quantum phenomenon. Of course, the properties of the atoms are determined by the laws of quantum mechanics and the quantum nature of their constituent particles, the electrons, and the atomic nucleus. Electrons do not seem to have substructure, but the atomic nucleus does—it being composed of neutrons and protons. And protons and neutrons appear to be made of more fundamental entities known as quarks. The properties and interactions of all of these particles are described quantum-mechanically. We do not know yet whether quarks are further reducible.

Perhaps one should not equate modern science with reductionism, but it is fair to say that the reductionist's approach has undeniably led to the overwhelming success of the modern scientific enterprise. But there are those who disparage the reductionist's approach—often without offering a coherent alternative. Some years ago we encountered a *New York Times* book review wherein the reviewer, who was not a scientist, made a comment to the effect that "reductionism is sometimes over-used in science", as if reductionism was merely a choice of literary style. We do not wish to enter the reductionism/ anti-reductionism debate here; but we do recommend the essay "The limitless power of science", by the University of Oxford chemist P. W. Atkins,* for a passionate defense of reductionism.

Well, can quantum mechanics itself be reduced? Can the strangeness of the quantum world be explained in terms of some deeper, more fundamental theory? Can the measurement problem be understood from such a theory? The probabilistic nature of quantum mechanics is well illustrated by what happens to a single photon upon striking a 50:50 beam-splitter. There is a 50% chance of its being reflected, and a 50% chance of its being transmitted. But according to standard quantum mechanics, there is no way to predict the trajectory of a *given* photon. The path selected is probabilistic—objectively indeterminate. Of course, one possibility for a more fundamental theory is a local hidden variable theory, which attempts to restore determinism and locality in order to "complete" quantum

---

* Published in *Nature's Imagination: the Frontiers of Scientific Vision*, ed. Jon Cornwell (Oxford University Press, Oxford, 1995). See also Steven Weinberg, in R. Feynman and S. Weinberg, *Elementary Particles and the Laws of Physics: The 1986 Dirac Memorial Lectures* (Cambridge University Press, Cambridge, 1987).

mechanics. But these theories, in contrast with quantum mechanics, make predictions in disagreement with some experiments, as we discussed in Chapter 5. Furthermore, it is a little difficult to imagine that there could be a more fundamental theory that would somehow be less weird than quantum mechanics itself. Perhaps quantum mechanics is the end of the road for reductionism.

Standard quantum theory is highly successful in all its applications, from photons and atoms, to semiconductors, superconductors, quarks, and so on. Yet it would still be nice to understand what the theory is really trying to tell us about the nature of the world beyond specific predictions. Such concerns force us into the realm of philosophy, or, as some have dubbed it, *quantum metaphysics*: the study of the interpretation of quantum mechanics. With regard to the philosophical foundations of quantum mechanics, there is a wide range of opinions. It is safe to say that the issue of the "correct" interpretation is still contentious, and it is likely to be so in the foreseeable future. The Murray Gell-Mann quote above is a good indication that not everyone is satisfied with the Copenhagen interpretation with which we have framed our explanations throughout the preceding chapters. We have clung to that interpretation, not necessarily out of complete agreement, but because it offers a consistent guide to the interpretation of experiments and because it has served as a guide in designing the experiments in the first place. Copenhagen surely has its weaknesses, especially in regard to the measurement problem. On the other hand, while there is contention over interpretation by some physicists, the average practicing quantum mechanic goes about doing his or her work making predictions through the mathematical formalism of the theory, or doing experiments, without too much regard for quantum philosophy. In fact, many of them despise it. Such physicists belong to what has been called the SUAC (Shut Up And Calculate!) school of thought. This agnostic outlook is unassailable. Again, so far, all known experiments are in agreement with the predictions of quantum theory. No-one questions that the theory is correct in that sense. What is in question is the issue of what kind of picture quantum mechanics paints for the nature of the universe at the most fundamental level. Volumes have been written on this topic, and we have no intention here of describing all the possible interpretations that have been proposed. Indeed, numerous alternative interpretations of quantum mechanics have been put forward—dozens have been proposed—and many of these interpretations themselves have variants that differ only slightly from the others. In fact, we have even seen papers which discuss what are called "classes of Copenhagen interpretations". To make matters worse, different names for the same interpretation are used by different authors. Most of these alternative interpretations have had scant impact on mainstream scientific thinking, and we hasten to add that we have not counted those interpretations, if they can even be called that, that invoke some form of magical thinking, mysticism, or "quantum consciousness" (see section 8.5).

Rather than attempt a survey of all the possible interpretations, or even of the classes of interpretation, we shall focus on the two most prominent ones: the Copenhagen interpretation, and one alternative interpretation that has been gaining some support in recent years, the *many-worlds* interpretation (MWI), also known as the many universes interpretation. We shall also, again, discuss the issue of decoherence and the issue of the quantum/classical border.

The central issues of contention are: (1) that the measurement of some attribute of a quantum system generally involves the probabilistic collapse, or reduction,[†] of the state vector where the probabilistic nature of the process is inherent—that is, not reducible to a more fundamental, deterministic explanation, and (2) the problem of bringing a measurement to a completion, as just discussed in the previous chapter in connection with the von Neumann "catastrophe" of infinite regression—a problem generally known as the *measurement paradox*. These two issues are not independent of each other. In the previous chapter we encountered one possible resolution of the measurement paradox wherein a measurement comes to completion with the collapse of the state vector in the human mind. But physicists, and scientists in general, find such a notion repellent.

We address the alternative interpretation (MWI) to take quantum mechanics to its (allegedly) logical conclusion—but first we recapitulate Copenhagen.

[†] The use of the word "reduction" is in reference to reduction, or collapse, of the state vector during measurement, and should not be confused with the issue of "reductionism".

8.2 Copenhagen and its Discontents

I have learned a great thing, a very great thing, that all that philosophers have ever written is pure drivel!

NIELS BOHR TO JENS LINDHARD[‡]

There is a certain irony in this quote from Niels Bohr (probably made in jest), as he is widely thought to have made his most important contribution to physics through his philosophical contemplations on the meaning and interpretation of quantum mechanics. Bohr's own writings on the quantum, though certainly not drivel, do tend to be obscure and at times impenetrable.[§] Nevertheless, throughout this book we have explained the results of various experiments with an appeal to the Copenhagen interpretation of quantum theory—the interpretation that grew out of Bohr's institute in Copenhagen. The principle architects of this interpretation were Niels Bohr, who developed the principle of complementarity, and Werner Heisenberg, who developed the uncertainty, or indeterminacy, principle, upon which the complementarity principle is based. But there were important contributions of others outside the immediate orbit of Copenhagen: namely, Max Born (of the University of Göttingen), who proposed the probabilistic interpretation of the square of quantum amplitudes, and John von Neumann (of Princeton University), who developed the so-called Hilbert space mathematical formulation of quantum theory, and introduced the projection postulate. Although Bohr and Heisenberg tried to present a unified front on the Copenhagen interpretation of quantum mechanics, the two were not in complete agreement, especially over the notion of complementarity. Their differences of opinion were a little subtle, but Bohr was rather inflexible and had practically elevated complementarity to the level of a dogma; he even tried to apply it to fields very far outside quantum mechanics. Furthermore, he apparently was not open to the idea that a deeper or more complete theory than Copenhagen quantum mechanics might be possible. There arose around him a

---

[‡] Comment made the day after attending a meeting of philosophers, as quoted in *Niels Bohr's Times*, by Abraham Pais (Clarendon Press, Oxford, 1991) p. 421.

[§] For a fascinating account of Bohr's thought processes and his associated difficulties with writing, see Richard Rhodes, *The Making of the Atomic Bomb* (Simon and Schuster, New York, 1986), Chapter 3.

cadre of sycophants who propagated (or propagandized upon, some have said) the notion that all problems with regard to the interpretation of quantum mechanics had been solved by Bohr, and that there was really nothing more to be said (reread the Gell-Mann quote above). This had the unfortunate consequence that any physicist politically incorrect enough to commit the heresy of speaking against the Copenhagen interpretation could expect to be under relentless attack and ridicule by the members of the Copenhagen school.

A good example is the case of the German physicist H. D. Zeh of the University of Heidelberg, whose career was damaged when he became interested in the measurement problem in the late 1960s. His case was one of those discussed recently in a paper entitled "Quantum dissidents: Research on the foundations of quantum theory *circa* 1970", by Olival Freire, Jr., in the journal *Studies in the History and Philosophy of Modern Physics*. In those days Zeh's work amounted to the first steps towards what is now known as the *theory of decoherence* of which we spoke in Chapter 7. Also at Heidelberg was J. Hans D. Jensen, a former associate of Bohr, who told Zeh that "any further activities on that subject would end [his] career!" Through Jensen, Leon Rosenfeld, already mentioned in Chapter 5 (Bohr passed away in 1962), learned of Zeh's work and labeled it "a concentrate of the wildest nonsense". In a 1980 letter to John Archibald Wheeler of Princeton University, Zeh wrote (quoting from the Freire paper):

> I have always felt bitter about the way how Bohr's authority together with [Wolfgang] Pauli's sarcasm killed any discussion about the fundamental problems of the quantum [...] I expect that the Copenhagen interpretation will some time be called the greatest sophism in the history of science, but I would consider it a great injustice if—when some day a solution should be found—some people claim that "this is of course what Bohr always meant", only because he was sufficiently vague.

Because of the attacks by the members of the Copenhagen school, many years would pass before other voices, such as John Bell's, were heard.**

** Some years ago, the senior author of this book (CCG) gave a talk about Bell's inequalities, and in the audience was a retired professor who had once been a post-doctoral research associate at Bohr's institute. After the talk he informed the audience that there was nothing of importance in the Bell inequalities, and that Bohr had already solved all the problems of quantum mechanics.

Let us summarize here the Copenhagen interpretation. We will start with a quote from Werner Heisenberg:

> The Copenhagen interpretation of quantum theory starts from a paradox. Any experiment in physics, whether it refers to the phenomena of daily life or to atomic events, is to be described in terms of classical physics. The concepts of classical physics form the language by which we describe the arrangement of our experiments and state the results … Still the application of these concepts is limited by the relations of uncertainty. (From *Physics and Philosophy*)

The main points of the Copenhagen interpretation—sometimes called the *orthodox* Copenhagen interpretation—are as follows:

1. A state vector $|\psi\rangle$ completely specifies what is known about a quantum system.
2. A state vector specifies probabilities for the measurement of a physical quantity. The probabilities apply to individual particles or quantum systems.
3. A measurement of a physical quantity produces an unpredictable, abrupt collapse of the state vector.
4. The Principle of Complementarity, which states that the attributes of complimentary physical quantities are not well defined; they cannot simultaneously exist except within the limits specified by the Heisenberg uncertainty principle.
5. The apparatus for performing and recording a measurement must be such that the results of the measurement can be understood in classical terms—that is, the measuring apparatus must be a classical device (as per the Heisenberg quote above).

The last point is necessary to avoid the issues raised in the "paradox" of Schrödinger's cat, and to put in place a divide between the quantum and classical worlds. But there is an ambiguity here. In principle, the measuring device can also be treated quantum-mechanically, and there is really no guide as to where to make the cut between the quantum and classical descriptions of the measurement. This is one of the weaknesses of Copenhagen. The experiments on magnetic flux trapped in a superconducting ring containing a Josephson junction described in the previous chapter do seem to suggest that the divide between these two worlds can be on a relatively large scale for a system sufficiently isolated from other systems.

To remind us of the practical implications of these points, we reconsider the double-slit experiment with electrons as described in Chapter 2. The state vector—or actually the wave function in this case—describes a single electron. Remember that the electrons pass through the slits one at a time. When a single electron passes through the slits but has not yet impacted on the screen, its associated wave function is spread out over the entire screen. The uncertainty in the electron's position is quite large. But when it finally does land on the screen, we see just an illuminated dot, indicating that its position is now much more localized. Thus, the electron's wave function has been abruptly and discontinuously reduced. At any point over the screen, before the electron impacts, the wave function is a superposition of the wave functions (amplitudes) associated with the two paths by which the electrons can reach the point. Because we do not try to detect that path taken by any electrons, the experiment displays their wave-like nature, and had we tried to obtain which-path information, the interference pattern would not have appeared. This is complementarity at work: the wave and particle natures of electrons (and photons, and so on) cannot be displayed simultaneously; that is, they are mutually exclusive attributes. Another example of complementarity is that of position and velocity (or momentum). Sharp values of those two quantities cannot exist simultaneously. Again, the important distinction between quantum and classical notions is that in classical physics we can assume that a particle—say a baseball—always has a definite position and momentum, even if these values are unknown. In quantum physics, however, these quantities are objectively indeterminate in accordance with the meaning of the Heisenberg uncertainty principle, position and momentum being complementary "observables". Quantum theory is probabilistic, in that it is not possible to predict with certainty where any given electron will land on the screen. However, the probabilities are determined from adding the quantum amplitudes associated with all possible paths by which the particle can reach the point on the screen. Because these amplitudes can be negative as well as positive (even complex), the probabilities can be zero in some places and large in others; that is, they can interfere destructively and constructively. In describing the motion of bullets through double slits (think back to the Born machine gun scenario), the distribution of bullets on the screen is also probabilistic (they clump opposite the holes), but there is no interference because the bullets, being macroscopic, always have well-defined momenta and positions. This, in turn, means that in principle we can follow them through the slits without affecting their motion. Finally, the screen, our "measurement device", acts as a classical object in the following sense: Once an electron impacts on the screen, the screen (or CCD camera, and so on) records the impact by a flash of light, and thus the information about the impact is irreversibly placed into the classical domain. Of course, the inter-action of the electron with the atomic structure of the screen must be described quantum-mechanically, but the resultant dot of light on the screen is macroscopic. The same idea applies in resolving the Schrödinger cat "paradox" of the previous chapter in the context of Copenhagen: the Geiger counter, if it detects a radioactive decay, irreversibly amplifies that detection into a macroscopic signal and thereby brings about a collapse of the state vector onto that of a dead cat.

A few more words about state vector collapse are necessary here. In the orthodox version of the Copenhagen interpretation, the collapse of the state vector, the discontinuous change of the state vector upon a measurement, is not a physical process. Rather, it is understood to be a mathematical process. The state vector itself is not interpreted as a real physical entity, so it should not be surprising that its collapse is not a physical process. There are variations of the Copenhagen interpretation that assert that the state vector is real, and that its collapse is physical.

As mentioned earlier in this book, the Copenhagen interpretation is closely related to the philosophical school of thought known as *logical positivism*. Logical positivism is an extreme form of *instrumentalism*— the view that the purpose of scientific theories is not to provide explanations of the world, but rather to provide mathematical models that make predictions that we can then compare with experiments. Here is what John von Neumann had to say on this issue:

> The sciences do not try to explain, they hardly even try to interpret, they mainly make models. But a model is meant as a mathematical construct which, with the addition of certain verbal interpretation, describes phenomena. The justification of such a mathematical construct is solely and precisely that it is expected to work.

This austere view is hardly universal among scientists: most really *do* want explanations beyond the predictions of a mathematical model. To them, a theory—any theory—needs to do more than make predictions that can be tested by experiments. A theory should lead to a coherent mental picture of phenomena and to understanding at the deepest level possible. It is this desire to have a complete picture of atomic-level phenomena that has led some to propose alternatives to quantum theory, and others to propose alternative interpretations of standard quantum mechanics.

Alternatives to quantum mechanics are generally slight modifications or extensions of standard quantum mechanics. There have been several attempts at extensions which are designed to bridge the micro and macro worlds. The most famous of these extensions is probably that of G. C. Ghirardi, A. Rimini, and T. Weber (GRW), wherein a non-linear interaction is added to the standard theory which randomly and spontaneously brings about a collapse of the quantum state, but where the collapse time depends on whether the number of particles involved is microscopic or macrscopic. The model is constructed in such a way that for a microscopic system, without the interaction of a measurement device, the quantum state essentially never collapses (just as in the Copenhagen interpretation), but for a macroscopic system the quantum state collapses within 100 nanoseconds (10^{-11} of a second) without interaction with any measurement device. Experiments of the type discussed in Chapter 7 could, in principle, detect such effects, but if they occur they must compete with the environmental effects that result in decoherence. Superpositions of macroscopically distinct quantum states decohere very rapidly, and might mask the effect proposed in the GRW theory. It would be necessary then to perform similar experiments in systems, but in extreme isolation from the rest of the universe. Such experiments have yet to be performed. In any case, if state vector collapse is not physical, as orthodox Copenhagen asserts, such experiments may amount to a fool's errand. No experiment would ever detect the collapse, no matter how good its temporal resolution.

Local hidden variable theories do provide underlying explanations for at least some aspects of quantum mechanics but, as we have seen, they make predictions that are different from those of quantum mechanics, as is illustrated by Bell's theorem. Thus, experiments can decide between the two theories, and all experiments to date overwhelmingly support quantum mechanics. We want to make clear that local hidden variable theories are not different interpretations of standard quantum mechanics; they are different theories altogether. They are now excluded by experiments. On the other hand, there exist *non-local* hidden variable theories of the type produced by David Bohm, but these make no testable predictions to discriminate between such theories and standard quantum mechanics. And there is still the "problem" of such theories being non-local. Because Bohm's theory was non-local, it did not satisfy Einstein.

8.3 The Many-Worlds Interpretation (MWI)

This interpretation supposedly attempts to restore "common sense" to quantum mechanics, but it does so at a price: it introduces multiple universes—a notion that surely argues against the conceit that "common sense" has been restored. Quantum weirdness does not disappear; it just shows up in a way that is quite different than the way it does in the Copenhagen interpretation. (This is a problem for all interpretations: quantum weirdness never goes away; it just manifests itself in different ways.)

The general idea behind MWI is this: When a measurement is performed on a quantum system in a superposition state, the state vector *never* undergoes a collapse to a particular state within the superposition. Rather, all possible outcomes of the measurement actually occur, but in different, mutually exclusive universes, sometimes simply called *worlds*—hence the name, *many-worlds interpretation* (MWI). These worlds are not connected to each other, and observers in each of them cannot contact each other and will even be unaware that the others exist. There is nothing even inherently probabilistic in this interpretation, as all possible outcomes of a measurement do occur—though, as we have said, in different universes.

Recall, from the previous chapter, the problem of the infinite regression in connection with the business of Schrödinger's cat. One could, taking seriously the idea of a quantum measurement entering the mind of the observer, write

$$|\psi\rangle = \frac{1}{\sqrt{2}} [|\text{not decayed}\rangle_{\text{atom}} \cdots |\text{alive}\rangle_{\text{cat}} |\text{sees cat alive}\rangle_{\text{observer}}$$
$$+ |\text{decayed}\rangle_{\text{atom}} \cdots |\text{dead}\rangle_{\text{cat}} |\text{sees cat dead}\rangle_{\text{observer}}],$$

where the dots (...) refer to the intermediate states involving the states of the flask and the hammer, and so on. Here the observer becomes entangled with the atom and the cat. To see quantum interference effects between the live and dead states of a macroscopic cat—and we mean a *real* cat here, as in the superposition state $|a\rangle$ above—is essentially impossible, due to the vast number of particles (degrees of freedom) that embody the cat. Or if it were possible, the interference pattern involving all these particles would be immensely complicated. Just as complicated would be an attempt to create interference between any of the other states of different systems in the entangled state $|\psi\rangle$. One could no more see interference between the states of the observer who sees the live or dead cat any more than one could of the live and dead cat states. The MWI asserts that the different chains of states in $|\psi\rangle$, the one with $|not$ $decayed\rangle_{atom} \cdots |alive\rangle_{cat} |sees$ cat alive$\rangle_{observer}$ and the one with $|decayed\rangle_{atom} \cdots |dead\rangle_{cat} |sees$ cat dead$\rangle_{observer}$, are in different universes, that the universe has *branched*, and that observer in the branch where the cat is alive will be unaware of the observer in the branch where the cat is dead. The MWI restores realism to quantum mechanics in the sense that both branches are asserted to exist, but at the price of being in different, and mutually inaccessible, universes.

With this interpretation, and depending on the experiment, or perhaps a sequence of experiments, it is possible to have a multiplicity of branching, or splitting, universes. To see how this could come about, consider a single photon initially prepared in the polarization state $|+45°\rangle$. Using polarizers we perform measurements to determine the photon's H or V polarization, recalling that $|+45°\rangle = (|H\rangle + |V\rangle)/\sqrt{2}$. Now look at Figure 8.1. The two lines representing the H and V results are to be understood as being in different universes (but remember, there is only one photon!). If measurements to determine $\pm 45°$ polarization are performed in each of those universes, then because those states are superpositions of H and V photon states, we get further splittings, so that we now have four universes (but still only one photon). Repeating this sequence of measurements in each of the subsequent universes simply increases the number of universes; that is, we have a proliferating sequence of splitting universes, and thus the state vector never comes to a collapse. In this interpretation there is no divide between the

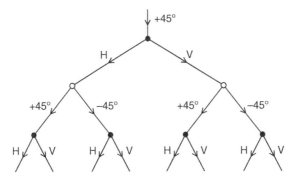

● = (H,V)measurement

○ = (±45°) measurement

Figure 8.1 In the many-worlds interpretation of quantum mechanics, a measurement produces all possible outcomes, each outcome being in a different world (or universe). We illustrate this with a sequence of measurements performed with polarizing filters on a state initially prepared as a superposition of horizontally and vertically polarized photons. Each line in the diagram represents a different universe.

quantum and classical worlds. The observer himself must be considered part of the system, and thus his state will also be split simultaneously into all these parallel universes.

Originally called the *relative state interpretation*, this interpretation was proposed in 1957 by Hugh Everett, in his doctoral thesis at Princeton University. Everett's ideas were attacked and ridiculed by the Copenhagen school, and thus were essentially ignored by most of the physics community for many years. Not until around 1970, through the efforts of Bryce Dewitt and others, did the MWI begin to gain a footing among all the other proposed interpretations. It does have a certain appeal for researchers in *quantum cosmology*, which attempts to treat the entire universe as a quantum system, for which there cannot be an external observer to perform measurements.

Though it may have provided much fodder for science fiction stories, the MWI has been called extravagant for obvious reasons. Does it really solve the measurement problem? It is hard to see how. Should you find yourself in the unfortunate situation of being in an airplane that is about to crash, you are not likely to be comforted by the notion that in some parallel universe you are safely in the same airplane landing on an airport runway. Aside from its extravagance, it is hard to see how one might regain the classical world from this interpretation, at least within its original formulation. However, it now appears that decoherence can come to the rescue and create the *appearance* of the collapse of the state vector, and thus also the *appearance* of a quantum/classical border. But there may be nothing about this unique to MWI: decoherence can explain the disappearance of all quantum aspects of the world (quantum superpositions, quantum correlations, and so on) on the large scale of the everyday macroscopic world without appealing to any particular interpretation.

Critics of the MWI often invoke William of Ockham's famous razor: *It is vain to do with more what one can do with fewer.* Supporters, though, say that Ockham's razor is properly applied to the number of *ideas* in the theory, not to the number of objects. And the MWI certainly does have its supporters—most notably the quantum information theorist David Deutsch, who defends this interpretation in his book *The Fabric of Reality* (see the bibliography at the end of this chapter). In fact, the MWI is allegedly becoming quite popular, even mainstream, but we seldom run into practicing physicists who accept this interpretation, though we suppose it is possible they could be using it in the privacy of their own homes. More typical, we believe, is the position of physicist R. F. Streater, who maintains on his website a list of Lost Causes in Theoretical Physics, in which the MWI is included (as are hidden variable theories). According to Streater, "There is nothing to the many-worlds theory. There are no theorems, conjectures, experimental predictions, or results of any sort, other than those of Hilbert space. It is not a cogent idea."[††]

The general problem with interpretations of quantum mechanics is that they *are* interpretations; they are interpretations of a set of laws and rules for calculations that make highly accurate predictions. Interpretations by their nature are not expected to make predictions that are any different from those of the standard theory. The debate

[††] Streater also lists as lost causes: quantum cosmology, hidden variable theories, and the prospect of converting Roger Penrose to the Copenhagen view.

over the many-worlds interpretation, or any of the other interpretations of quantum mechanics, is not likely to end soon.[‡‡]

8.4 Decoherence

Decoherence itself is sometimes considered an interpretation of quantum mechanics. As one might expect, there are several "decoherence"
interpretations. However, we prefer to think of decoherence as a
phenomenon—a phenomenon brought about by the interaction of
a quantum system with its environment as explained at the end of the
previous chapter, the idea being that a quantum system is never truly
isolated from interactions with its environment, and that these interactions cause quantum superpositions to be destroyed and trans-
formed into statistical mixtures. One can think of the interaction of
a small system with its environment as a kind of continuous measurement being performed on that system. What is happening is that
the states of the small system become entangled with the very large
number of quantum states available to the large system consisting of
the rest of the universe and the small system itself. This has the effect
of washing out the coherences of the small system. Now, if a system
consisting of a small number of particles is isolated from the rest of
the universe, then superpositions of the states and entangled states of
these particles are relatively easy to maintain, and in fact, such is
routinely done these days in systems of small numbers of atoms and/
or photons. But the larger and more complicated the system (think of
Schrödinger's cat), the more difficult it is to isolate it from the rest of
the universe. In fact, the larger the system the more rapid should its
superposition states decohere into mixtures. However, as a matter of
principle, it ought to be possible to produce superpositions and
entanglements of large systems, depending on how clever the experimenter is in keeping the system isolated for a sufficiently long time in
order to perform the measurements required to demonstrate quantum coherence on a large-scale system. It has even been shown that
entangled states of two or more large-scale systems, if sufficiently
protected from the effects of decoherence, can be used to falsify local

---

[‡‡] For other opinions about the MWI, see the website http://plato.stanford.edu/entries/
qm-manyworlds/.

hidden variable theories in experiments related to Bell's theorem, as discussed in Chapter 5.

From the point of view of decoherence, as we said at the end of the previous chapter, there really is no divide between the quantum and classical worlds. When we study any system that has become entangled with its environment through interactions with it, measurements cannot be performed on this much larger system to reveal overall superposition and entanglement. Experiments always examine only a small part of the universe and ignore the rest. But that does not mean that the rest of the universe (the environment) ignores the system. If the system and its environment interact for a long enough time (a time which in practice can be very short), then the ensuing entanglement between the small system of interest and the large system that comprises its environment cause a complete loss of coherence in the former. Thus, ironically, quantum entanglement with a large system renders the smaller into a system lacking quantum coherence that effectively renders the latter into a system that appears classical. If one increases the size of the smaller system (by, say, having more particles), the only effect is that loss of quantum coherence (decoherence) of that system proceeds much faster, and that is why we do not encounter the weird Schrödinger-cat-like states in the everyday world. From the point of view of decoherence, there is no quantum/classical divide.

Decoherence is the mortal enemy of the quantum computer. To build a large-scale quantum computer will require thousands of particles to be maintained in quantum coherences for relatively long periods of time. The difficulty of isolating such a system from the effects of decoherence cannot be overstated. For this reason, we do not expect to see a large-scale quantum computer for some time to come.

8.5 Quantum Consciousness?

We have already encountered the notion, proposed by Wigner and others, that the collapse of the state vector may ultimately be brought about by the human mind, or at least some mind (perhaps the mind of a cat). That is one way that consciousness can be perceived as having entered the quantum world. But as we have emphasized throughout this book, the experimenter—otherwise known as the observer—is free to make a conscious choice between different kinds of experiments to perform; that is, free to choose different kinds of experiments which give different kinds of results. Referring again to the double-slit experiment with electrons, if one chooses to obtain which-path information by closing off one of the slits, then the particle nature of electrons is manifested. Choosing not to obtain which-path information by leaving both slits open results in quantum interference, the manifestation of the wave-like nature of electrons, the particle and wave properties being complementary. And, recall that it does not even matter when the decision is made as to which properties will be displayed (at least prior to detection), as is shown by the delayed-choice experiment. Because the observer—a conscious entity—makes choices over the type of experiment to be performed, and therefore of the kinds of results that will be obtained, consciousness does have a role in the quantum world—at least in well-defined experimental situations. One can, of course, imagine an automated experiment where the choices are made by a computer program, though this changes nothing, as one could simply say that the choices were made during the writing of the computer program that runs the experiment. To avoid that complication, one could instead use a quantum-mechanical random-number generator of the type discussed in Chapter 3 (a single photon entering a beam-splitter), whose outcomes randomly decide what kind of experiment to perform. If the photon is reflected, one type is performed; if transmitted, the other type.

Unfortunately, the fact that one can choose different kinds of experiments to obtain different kinds of results has led to the widely held belief, at least in some segments of the general public, that, through quantum mechanics, the entire universe is *observer-created*. Such claims are hyperbolic at best, and have the smell of the philosophy of Hegel.[§§] For some time now, various quantum carpetbaggers, with little or no expertise in quantum physics, have been making exaggerated claims of alleged connections between human consciousness and quantum mechanics,[***] and recently a film has appeared to

[§§] See the essay "Philosophy and Politics" in *Unpopular Essays*, by Bertrand Russell (Simon and Shuster, New York, 1950). Hegel: " . . . all reality is thought."

[***] Particularly egregious examples, in our opinion, are any of the books written by Fred Alan Wolf (*The Dreaming Universe, The Yoga of Time Travel, The Spiritual Universe*, and so on). These books are not taken seriously by the scientific community.

the same effect.[†††] There has even been a claim that quantum effects have direct medicinal benefits[‡‡‡]—a notion popular among practitioners of homeopathy. Such claims are unwarranted.

On the other hand, it seems legitimate to ask: What does quantum mechanics have to do, if anything, with the *origin* of consciousness? Consciousness, whatever one chooses to believe about it, may ultimately turn out to be a quantum phenomenon—a view strongly supported by eminent physicist Roger Penrose (who, by the way, also suspects that gravity has something to do with quantum state reduction). Consciousness certainly involves chemical and electrical processes taking place in the brain. In fact, we can easily alter our state of consciousness through the introduction of chemicals, as in anesthetics, or by physical injury. And, yes, chemical reactions are fundamentally quantum-mechanical, but so are the reactions that take place in the internal combustion engine of your car, though you do not normally think of your car's engine as being a quantum-mechanical device. The weird features exhibited by superpositions and entanglements—that is, *quantumness*, of quantum coherence—are not *obviously* manifested by mental activity. On the other hand, there is evidence that quantum coherence effects play a role in some biological processes. For example, there is some recent evidence for wave-like energy transfer via quantum coherence in photosynthesis. But it is a big leap to assume that quantum coherences can exist in the brain. Even should that eventually prove to be the case, it would not follow that that consciousness could control such coherences and somehow use them to effect the external world in a manner that most of the quantum consciousness advocates would like us to believe. Their claims must be viewed with utmost skepticism and due regard for the principles of the scientific method.

In summary, we urge the reader to beware of exaggerated claims of quantum effects in the world of everyday life—especially those involving allegations of quantum consciousness. Quantum mechanics is weird, but not *that* weird.

[†††] "What the Bleep Do You Know?" This movie is a mix of a small bit of science fact and many wild, unsubstantiated claims.
[‡‡‡] For example, *Quantum Healing*, by Deepak Chopra.

8.6 The Mystery Remains

We dance round in a ring and suppose,
But the Secret sits in the middle and knows.

ROBERT FROST, From his poem *The Secret Sits*

We have sampled only the surface of the vast literature on the interpretation of quantum mechanics, its very vastness testament to the unsettling weirdness of the quantum world. When one examines these writings on the problem, one is struck by the fact that in just about all, if, indeed, not all, of the possible interpretations, the fundamental mystery never disappears. There appears to be a "conservation of quantum weirdness" law operating with respect to interpretations of quantum mechanics. The weirdness may disappear in one place, but it appears somewhere else. The many-worlds interpretation does not really solve the mystery of the quantum, unless one thinks that an unseen infinity of splitting universes is not a mystery in its own right. One *can* apparently make the mysteries disappear by abandoning standard quantum mechanics altogether and resorting to local hidden variable theories, but as we have discussed, these do not stand up to experiments. Attempts to modify those kinds of theories in an attempt to put them in agreement with laboratory results appear to be strained, and sometimes invoke mechanisms that are just as strange as those of the standard theory.

The mystery of the quantum remains. As Richard Feynman has written in his book *The Character of Physical Law*:

> I think I can safely say that nobody understands quantum mechanics...Do not keep saying to yourself, if you can possibly avoid it, "but how can it be like that?" because you will get "down the drain", into a blind alley from which nobody has yet escaped. Nobody knows how it can be like that.

Bibliography

Ball P., "Physics of life: The dawn of quantum biology," Nature (News Feature) Vol. 474, page 272 (2011).

Camilleri K., "A history of entanglement: Decoherence and the interpretation problem," Studies in History and Philosophy of Modern Physics Vol. 40, page 290.

Deutsch D., *The Fabric of Reality*, Penguin Books, 1997.

DeWitt B. S., and Graham N., Editors, *The Many-Worlds Interpretation of Quantum Mechanics*, Princeton University Press, 1973.

DeWitt B. S., "Quantum mechanics and reality," Physics Today, February, 1970, page 35.

Engel G. S., Calhoun T. R., Read E. L., Ahn T.-K., Mančal T., Cheng Y.-C., Blankenship R. E., and Fleming G. R., "Evidence of wavelike energy transfer through quantum coherence in photosynthetic systems," Nature Vol. 446, page 782 (2007).

Freire O., Jr. "Quantum dissidents: Research on the foundations of quantum theory circa 1970," Studies in the History and Philosophy of Modern Physics, Vol. 40. Page 280 (2009).

Ghirardi G. C., Rimini A., and Weber T., "Unified dynamics for microscopic and macroscopic systems," Physical Review D Vol. 34, page 470 (1986).

Heisenberg W., *Physics and Philosophy: The Revolution in Modern Science*, Harper, 1958.

Osnaghi S., Freitas F., and Freire O., Jr. "The origin of the Everettian heresy," Studies in History and Philosophy of Modern Physics Vol. 40, page 97.

Pais A., "*Niels Bohr's Times*", *in Physics, Philosophy, and Polity*, Oxford University Press, 1991.

Penrose R., *The Emperor's New Mind*, Oxford University Press, 1989.

Penrose R., *Shadows of the Mind*, Oxford University Press, 1994.

Wigner E. P., *Symmetries and Reflections, Scientific Essays*, MIT Press, 1970.

APPENDIX A

A Quantum Mechanics Timeline

The following is an historical outline of the developments of quantum mechanics particularly relevant to the text. It begins in the year 1900 and continues to more or less the present. It is not meant to be exhaustive.

1900   Max Planck explains the spectral distribution of a black body by introducing the idea that the energy exchanged between the walls of a cavity and the radiation field inside is quantized.

1905   Albert Einstein uses Planck's idea of quantized energy and introduces the concept of light quanta (now called photons) to explain the photoelectric effect. The photon is understood as a "particle" of light—light generally understood to be wave-like.

1907   Ernest Rutherford discovers the structure of the atom, that all the positive charge of the atom is located in a small, massive nucleus. He scatters alpha particles through gold foil, and finds that while most pass right through, some are scattered straight backwards, which he explains as being due to head-on collisions.

1913   Niels Bohr uses the Rutherford model of the nuclear atom, and introduces the idea that the energies of the electrons around the nucleus are quantized to explain the origin of spectral lines for hydrogen. Bohr's quantum theory is *ad hoc*, and does not explain the intensities of spectral lines.

1921   Otto Stern and Walter Gerlach discover "space quantization", wherein a beam of silver atoms is split into two beams upon passage through a non-uniform magnetic field. The effect is later understood as being due to the magnetic moment of a single-valence electron of the silver atoms, where the electron spin is quantized in magnitude and direction—only two directions, "up" or "down".

1923   Louis de Broglie suggests that as a wave-like phenomena such as light can have a particle-like nature as shown by Einstein, perhaps particles such as electrons can have a wave-like nature.

1925   Wolfgang Pauli introduces the Pauli exclusion principle, which states that no two electrons in an atom can have the same set of quantum numbers. The principle explains the electron structure of complex atoms.

1925   Werner Heisenberg invents a new form of quantum theory, quantum mechanics, which is capable of explaining spectral line intensities. His theory is later called matrix mechanics.

1925–26 Erwin Schrödinger, working with de Broglie's notion of a wave-like nature of particles, discovers another form of quantum mechanics: wave mechanics. He later shows that wave mechanics and matrix mechanics are equivalent theories.

1927   Clinton J. Davisson and Lester A. Germer, at Bell Laboratories in New York City, observe electron diffraction, demonstrating the wave-like nature of electrons.

1927   Niels Bohr introduces the notion of complementarity into quantum mechanics.

1932   Jon von Neumann introduces the Hilbert space mathematical formalism approach to quantum mechanics, and also introduces the projection postulate.

1935   Einstein, Podolski, and Rosen publish their attack on the Copenhagen interpretation of quantum mechanics, insisting that the theory provided an incomplete description of the quantum world. They implicitly introduce the notion of an entangled state.

1935	Erwin Schrödinger introduces the word "entanglement" into quantum mechanics to describe multi-particle states that cannot be factorized. He notes that entanglement is what truly characterizes the structure of quantum mechanics. In addressing the issue of quantum measurement he introduces the paradigm of Schrödinger's cat, intended as a burlesque of the Copenhagen interpretation of measurement.
1952	David Bohm introduces a hidden variable theory formulation of quantum mechanics, but the theory is non-local and is formally equivalent to standard quantum mechanics. Einstein is unimpressed by this theory, commenting that Bohm "got his results cheap".
1964	John Bell shows that the local hidden variable theories can, in some cases, predict results different from standard quantum mechanics. He devises a way to test both theories through the use of an inequality satisfied by local hidden variable theories, but violated by standard quantum mechanics.
1972	John Clauser and Stuart Freedman perform the first experimental test of Bell's inequality, using polarized photons from an atomic source. The experiment tentatively showed violations of Bell's inequality, thus apparently ruling out local hidden variable theories. There were, though, some loopholes in the experiment.
1982	Alain Aspect and collaborators close most of the loopholes of the Clauser–Freedman experiment (by using ultra-fast switching), and provide the strongest support yet for standard quantum mechanics.
Mid-1980s through to the present	Various experimenters, Lenard Mandel, Paul Kwiat, and others, use photon pairs obtained from down-conversion processes, and obtain violations of Bell inequalities by as much as twenty standard deviations.
1984	C. H. Bennett and G. Brassard propose a protocol for quantum key distribution (quantum cryptography).

1985   Gérard Roger and collaborators demonstrate single-photon interference.

1986   Two groups—one headed by Carol Alley at the University of Maryland, and another headed by Herbert Walther at the Max Planck Institute in Munich—perform delayed-choice experiments using attenuated laser light.

1987   C. K. Hong, Z. Y. Ou, and L. Mandel perform an experiment showing that when single photons simultaneously fall on opposite sides of a beam-splitter, both photons emerge in the same path.

1989   Daniel Greenberger, Michael Horne, and Anton Zeilinger propose another form of Bell's theorem using entangled states of three or more particles. They show that with such states (GHZ states), local hidden variable theories can be falsified with a single run of an experiment.

1992   Paul Kwiat and collaborators experimentally demonstrate the quantum eraser.

1993   A. M. Steinberg and collaborators experimentally measure the tunneling time of a single photon, and find an apparent tunneling velocity of about 1.7 times the speed of light.

1993   L. Hardy proposes a test of local realism against quantum mechanics, using two entangled states but without using Bell's inequality.

1994   P. W. Shor presents a quantum-mechanical-based procedure for finding the prime factors of large integers in a short time, if a quantum computer were to exist.

1995   Paul Kwiat and collaborators experimentally demonstrate interaction-free measurement.

1995   L. Mandel's group experimentally demonstrates the Hardy proposal, and finds agreement with quantum mechanics.

1997–98   Two groups—one headed by Anton Zeilinger, in Innsbruck at the time, and another headed by Francesco De Martini in Rome—experimentally demonstrate quantum teleportation.

2000 Anton Zeilinger and collaborators perform a test of local hidden variable theories using three photons in an entangled state. Local hidden variable theories fail the test.

2000 Two groups—J. R. Friedman and collaborators at the State University of New York, Stony Brook, and C. H. van der Wal and collaborators at Delft University of Technology—provide experimental evidence that a SQUID can be placed into a superposition of two magnetic-flux states.

2006 A group headed by Philippe Grangier and Alain Aspect perform an almost ideal version of the delayed-choice experiment with single photons.

2007 Anton Zeilinger's group performs an experiment, based on an idea of Anthony Leggett, to test non-local realism. The test shows that non-local realistic theories are incompatible with standard quantum mechanics.

2007 The Japanese collaboration headed by Y. Yamamoto performs an experiment in quantum key distribution, using single photons and efficient superconducting photon detectors, over a distance of 200 km.

2009 A collaboration between a group at the University of Geneva and Corning Incorporated performs high-rate quantum key distribution over 250 km with ultra-low-loss optical fibers.

2009 A group headed by Christopher Monroe at the University of Maryland performs an experiment teleporting the quantum state of a trapped ion to another trapped ion separated from the first by a distance of about 1 meter.

Bibliography

The following discuss many of the topics of this book along historical lines:

B. Hoffman, *The Strange Story of the Quantum*, 2nd edition, Dover Publications, 1959.

V. Guillemn, *The Story of Quantum Mechanics*, Dover Publications, 2003.

G. Gamov, *The Thirty Years that Shook Physics*, Dover Publications, 1885.

M. Jammer, *The Conceptual Development of Quantum Mechanics*, McGraw-Hill, 1966.

M. Jammer, *The Philosophy of Quantum Mechanics*, John Wiley & Sons, 1974.

H. Kragh, *Quantum Generations, A History of Physics in the Twentieth Century*, Princeton University Press, 1999.

A. Whitaker, *Einstein, Bohr, and the Quantum Dilemma*, Cambridge University Press, 2006.

D. Lindley, *Uncertainty: Einstein, Heisenberg, Bohr, and the Struggle for the Soul of Science*, Anchor Books, 2008.

E. Segrè, *From X-rays to Quarks, Modern Physicists and Their Discoveries*, Dover, 2007.

G. Segrè, *Faust in Copenhagen: A Struggle for the Soul of Physics*, Penguin Book, 2007.

L. Gilda, *The Age of Entanglement*, Vintage, 2008.

M. Kumar, *Quantum: Einstein, Bohr, and the Great Debate About the Nature of Reality*, Norton, 2008.

APPENDIX B

Quantum Mechanics Experiments for Undergraduates

Over the past decade, a group of dedicated physics professors has spearheaded the development of experiments for performing fundamental tests of quantum mechanics in the undergraduate physics laboratory. Most of these experiments are optical, and several of the experiments described in this book are among those that can now be performed by undergraduates. These include the demonstration of the existence of single photons (single photon falling on a beam splitter), single-photon interference, violations of Bell's inequalities using entangled photons, the quantum eraser with entangled photons, and the Hardy–Jordan test of local realism. For an overview with references and links to groups developing such experiments, see the website maintained by Professor Mark Beck of Whitman College: Modern Undergraduate Quantum Mechanics Experiments, http://people.whitman.edu/~beckmk/QM/. Also see the website of Professor Enrique Galvez of Colgate University (www.colgate.edu/facultysearch/FacultyDirectory/Egalvez for additional resources on quantum mechanics experiments for undergraduates.

General Bibliography

Andrade J., e Silva and G. Lochak, *Quanta*, McGraw-Hill, 1969.

Audretsch J., *Entangled World*, Wiley-VCH, 2006.

Baggott J., *The Meaning of Quantum Theory*, Oxford University Press, 1992.

Born M., *The Born–Einstein Letters 1916–1955*, Macmillan, 1971.

Feynman R., *The Character of Physical Law*, MIT Press, 1967.

Ford K. W., *The Quantum World*, Harvard University Press, 2004.

Ghirardi G., *Sneaking a Look at God's Cards*, Princeton University Press, 2005.

Greenstein G., and Zajonc A. H., *The Quantum Challenge*, 2nd edn., Jones and Bartlett, 2006.

Heisenberg W., *Physics and Philosophy*, Harper & Brothers, 1958

Hughes R. I. G., *The Structure and Interpretation of Quantum Mechanics*, Harvard University Press, 1989.

Laloë F., "Do we really understand quantum mechanics? Strange correlations, paradoxes, and theorems", *American Journal of Physics* 69 (2001), 655.

Lederman L., and Hill C. T., *Quantum Physics for Poets*, Prometheus Books, 2011.

Lindley D., *Where Does All the Weirdness Go?*, Basic Books, 1996.

Onnès R., *The Interpretation of Quantum Mechanics*, Princeton University Press, 1994.

Onnès R., *Quantum Philosophy*, Princeton University Press, 1999.

Onnès R., *Understanding Quantum Mechanics*, Princeton University Press, 1999.

Pagels H. R., *The Cosmic Code*, Simon and Schuster, 1982.

Rae A., *Quantum Physics: Illusion or Reality?*, 2nd edn.,Cambridge University Press, 2004.

Rae A., *Quantum Physics: A Beginner's Guide*, Oneworld Publications, 2005.

Shimony A., "The Reality of the Quantum World", *Scientific American*, January 1988, p. 46.

Shimony A., "Conceptual foundations of quantum mechanics", in *The New Physics*, ed. Paul Davies, Cambridge University Press, 1989, p. 373.

Styer A., *The Strange World of Quantum Mechanics*, Cambridge University Press, 2000.

Treiman A., *The Odd Quantum*, Princeton University Press, 1999.

Whitaker A., *Einstein, Bohr, and the Quantum Dilemma*, 2nd edn., Cambridge University Press, 2006.

Whitaker M. A. B., "Theory and experiment in the foundations of quantum mechanics", *Progress in Quantum Electronics* 24 (2000), 1.

Zeilinger A., *Dance of the Photons: From Einstein to Quantum Teleportation*, Farra, Straus and Giroux, 2010.

Index

Page numbers in *italic* indicate figures and tables

Printed and bound by CPI Group (UK) Ltd, Croydon, CR0 4YY